Transboundary Water Resources: A Foundation for Regional Stability in Central Asia

NATO Science for Peace and Security Series

This Series presents the results of scientific meetings supported under the NATO Programme: Science for Peace and Security (SPS).

The NATO SPS Programme supports meetings in the following Key Priority areas: (1) Defence Against Terrorism; (2) Countering other Threats to Security and (3) NATO, Partner and Mediterranean Dialogue Country Priorities. The types of meeting supported are generally "Advanced Study Institutes" and "Advanced Research Workshops". The NATO SPS Series collects together the results of these meetings. The meetings are co-organized by scientists from NATO countries and scientists from NATO's "Partner" or "Mediterranean Dialogue" countries. The observations and recommendations made at the meetings, as well as the contents of the volumes in the Series, reflect those of participants and contributors only; they should not necessarily be regarded as reflecting NATO views or policy.

Advanced Study Institutes (ASI) are high-level tutorial courses intended to convey the latest developments in a subject to an advanced-level audience

Advanced Research Workshops (ARW) are expert meetings where an intense but informal exchange of views at the frontiers of a subject aims at identifying directions for future action

Following a transformation of the programme in 2006 the Series has been re-named and re-organised. Recent volumes on topics not related to security, which result from meetings supported under the programme earlier, may be found in the NATO Science Series.

The Series is published by IOS Press, Amsterdam, and Springer, Dordrecht, in conjunction with the NATO Public Diplomacy Division.

Sub-Series

A.	Chemistry and Biology	Springer
B.	Physics and Biophysics	Springer
C.	Environmental Security	Springer
D.	Information and Communication Security	IOS Press
E.	Human and Societal Dynamics	IOS Press

http://www.nato.int/science
http://www.springer.com
http://www.iospress.nl

Series C: Environmental Security

Transboundary Water Resources: A Foundation for Regional Stability in Central Asia

Edited by

John E. Moerlins
Florida State University,
U.S.A.

Mikhail K. Khankhasayev
Florida State University,
U.S.A.

Steven F. Leitman
Florida State University,
U.S.A.

and

Ernazar J. Makhmudov
Institute of Water Problems,
Uzbek Academy of Sciences,
Uzbekistan

Springer

Published in cooperation with NATO Public Diplomacy Division

Proceedings of the NATO Advanced Research Workshop on
Facilitating Regional Security in Central Asia through improved Management
of Transboundary Water Basin Resources
Almaty, Kazakhstan
20–22 June 2006

A C.I.P. Catalogue record for this book is available from the Library of Congress.

ISBN 978-1-4020-6735-8 (PB)
ISBN 978-1-4020-6734-1 (HB)
ISBN 978-1-4020-6736-5 (e-book)

Published by Springer,
P.O. Box 17, 3300 AA Dordrecht, The Netherlands.

www.springer.com

Printed on acid-free paper

All Rights Reserved
© 2008 Springer Science+Business Media B.V.
No part of this work may be reproduced, stored in a retrieval system, or transmitted
in any form or by any means, electronic, mechanical, photocopying, microfilming,
recording or otherwise, without written permission from the Publisher, with the exception
of any material supplied specifically for the purpose of being entered
and executed on a computer system, for exclusive use by the purchaser of the work.

CONTENTS

Preface ix

Introduction to NATO Advanced Research Workshop
on Transboundary Water Management in Central Asia 1
John E. Moerlins and Ernazar J. Makhmudov

Part I: Transboundary Water Management Problems in the Central Asia Region

1. Problems of Water Resource Management in Central Asia 11
   *Ernazar J. Makhmudov, Ilhomjon E. Makhmudov,
   and Lenzi Z. Sherfedinov*

2. Communicating the Issues of the Aral Sea Basin
   Long-term Vision for the Aral Sea Basin
   the Aral Sea Basin Management Model (ASB-MM) 29
   Joop De Schutter

3. Water Ecosystems of Central Asia: Important Factors
   Affecting the Environmental & Social Prosperity of the Region 43
   *Yessekin Bulat, Burlibayev Malik, Medvedeva Nina,
   and Stafin Sanzhar*

4. NATO/CCMS Pilot Study Meeting on Transboundary
   Water Management Issues in the United States & Central
   Asia: Problem Definition, Regulation, and Management 65
   Mikhail Khankhasayev and Steven Leitman

Part II: Management of Transboundary Water Resources in Central Asia and Caucasus Region

5. Integrated Management of Transboundary Water
   Resources in the Aral Sea Basin 79
   Nariman Kipshakbaev

CONTENTS

6. Transition to IWRM in Lowlands of the Amu Darya and the Syr Darya Rivers 87
   Victor A. Dukhovny and Mikhail G. Horst

7. Improvement of Water Resources Management in the Aral Sea Basin: Subbasin of the Amu Darya River in its Middle Reach 105
   Kurbangeldy Ballyev

8. Integrated Water Resource Management of Transboundary Chu and Talas River Basins 123
   Elene Rodina, Anna Masyutenko, and Sergei Krivoruchko

9. Chu-Talas Activities 131
   Lea Bure

10. Mechanisms for Improvement of Transboundary Water Resources Management in Central Asia 141
    Dushen M. Mamatkanov

11. Science for Peace: Monitoring Water Quality and Quantity in the Kura–Araks Basin of the South Caucasus 153
    Michael E. Campana, Berrin Basak Vener, Nodar P. Kekelidze, Bahruz Suleymanov and Armen Saghatelyan

12. Hazardous Pollutant Database for Kura–Araks Water Quality Management 171
    Bahruz Suleymanov, Majid Ahmedov, Famil Humbatov, and Navai Ibadov

13. On Development of GIS-Based Drinking Water Quality Assessment Tool for the Aral Sea Area 183
    Dilorom Fayzieva, Elena Kamilova, and Bakhtiyar Nurtaev

Part III: Legal, Technical and Institutional Aspects of Transboundary Water Management

14. Lessons Learned from Transboundary Management Efforts in the Apalachicola–Chattahoochee–Flint Basin, USA 195
    Steve Leitman

15. Determining Equitable Utilization of Transboundary Water Resources: Lessons from the United States Supreme Court 209
    George William Sherk

16. Improving Transboundary River Basin Management by Integrating Environmental Flow Considerations 223
    Karin M. Krchnak

17. Transboundary Aquifers as Key Component of Integrated Water Resource Management in Central Asia 243
    Ken Howard and Anne Griffith

18. Integrated Water Management 263
    J. Staes, H. Backx, and P. Meire

Part IV: Workshop Conclusions and Recommendations

ARW Conclusions and Recommendations 305

Subject Index 311

PREFACE

The hydrologic cycle on Earth supports an abundance of life, including human life. Freshwater ecosystems are supported by the hydrologic cycle, providing immeasurable services and benefits to humans, ranging from food and water purification to spiritual renewal. Despite growing awareness of the importance of healthy freshwater ecosystems, human actions continue to degrade the freshwater ecosystems upon which we depend. Even with the policy movement toward integrated water resource management, the integration of ecosystem considerations in water management remains largely neglectted. Since the early 1960s, mismanagement of water resources has plagued the Aral Sea Basin. The problem became more complex in 1991 when the Soviet Union collapsed and the Aral Sea Basin became a transboundary water body. Overnight, the development of sustainable and equitable water management practices became the shared responsibility of five sovereign nations each with conflicting needs, goals, and priorities.

In response to this growing problem, on 20–22 June 2006, a NATO-sponsored Advanced Research Workshop (ARW) was organized in Almaty, Kazakhstan on Facilitating Regional Security in Central Asia through Improved Management of Transboundary Water Basin Resources. The co-organizers of this workshop were the Institute for International Cooperative Environmental Research (IICER) at Florida State University (FSU) and the Institute of Water Problems (IWP) of the Academy of Sciences of the Republic of Uzbekistan. The precursor to this NATO/ARW was the NATO/CCMS Pilot Study on Environmental Decision-Making for Sustainable Development in Central Asia that was also organized and administered by the IICER at FSU.

The final meeting of the pilot study was conducted in Florida in March of 2005. This meeting, represented the culmination of the five years of the work of this pilot study, and focused on transboundary water issues with a focus on the southeastern United States (i.e., Florida, Alabama, and Georgia) and how this experience may be applied to the Central Asian region. Based on the results of this final pilot study meeting, a unanimous conclusion was reached by the meeting participants to further investigate this fundamental and important environmental problem of the Central Asian region: shared fresh water resources. This conclusion led to the development of a proposal to conduct this NATO Advanced Research Workshop. This three-day workshop included participation by approximately 30 environmental professionals from 12 countries. Current issues facing the Central Asian region related to Transboundary Water Management were discussed, including specific "case studies" impacting the individual republics. A section on Workshop Conclusions and Recommendations is provided based on the discussions at the workshop.

We wish to express our professional and personal appreciation to the individuals whose expertise, dedication, and hard work made the workshop a success. We especially would like to acknowledge Mr. Steven Leitman for his important impact on the organization and implementation of this NATO Workshop, and serving as a coeditor of these proceedings. We express our appreciation to Dr. Roy Herndon for his efforts and support in organizing and conducting this workshop and the staff of the Institute for International Cooperative Environmental Research at Florida State University. We would like to thank the Center for Basic and Ecological Research (Almaty, Kazakhstan) and, especially Dr. Nurgali Takibaev, a Director, for assistance in organizing this workshop in Almaty. The organizers would like to recognize the efforts of Mr. Norbert Barszczewski for his organizational skills and for preparation of this manuscript. The Organizing Committee would like to thank the following individuals for their significant contributions to the organization and execution of this Advanced Research Workshop: Elmira Baisetova, Paul Winkel, and Randie Denker.

NATO ARW CODIRECTORS

MR. JOHN E. MOERLINS AND DR. MIKHAIL KHANKHASAYEV

Institute for International Cooperative Environmental Research, Florida State University

PROFESSOR ERNAZAR MAKHMUDOV

Institute of Water Problems, Academy of Sciences of the Republic of Uzbekistan

INTRODUCTION TO NATO ADVANCED RESEARCH WORKSHOP ON TRANSBOUNDARY WATER MANAGEMENT IN CENTRAL ASIA

JOHN E. MOERLINS
Institute for International Cooperative Environmental Research (IICER), Florida State University, Tallahassee, Florida, USA

ERNAZAR J. MAKHMUDOV
Institute of Water Problems (IWP), Uzbek Academy of Sciences, Tashkent, Uzbekistan

Abstract: This paper summarizes the activities and results of a three-day NATO Advanced Research Workshop on transboundary water management issues in Central Asia. This NATO ARW was conducted in Almaty, Kazakhstan on 20–22 June 2006, involving participation by water management experts from the Central Asian region and from NATO countries.

Keywords: NATO Advanced Research Workshop, transboundary water management, conflict avoidance, Central Asia, environmental decision-making, legal/technical/institutional determinants of water issues

1. Introduction

On 20–22 June 2006, a NATO-sponsored Advanced Research Workshop (ARW) was organized in Almaty, Kazakhstan entitled *Facilitating Regional Security in Central Asia through Improved Management of Transboundary Water Basin Resources*. The co-organizers of this workshop were the Institute for International Cooperative Environmental Research (IICER) at Florida State

University (FSU) and the Institute of Water Problems (IWP) of the Uzbek Academy of Sciences.

The precursor to this ARW was the five-year NATO/CCMS Pilot Study on *Environmental Decision-Making for Sustainable Development in Central Asia* that was organized and administered by the IICER at FSU. The final meeting of the pilot study was conducted in Florida in March 2005. The final meeting of this pilot study in 2005 was focused on transboundary water issues utilizing, as an example, an on-going water conflict in the southeastern United States (i.e., Florida, Alabama, and Georgia). Based on the results of this final pilot study meeting, a unanimous conclusion reached by the meeting participants was to further investigate this fundamental and important environmental problem of the Central Asian region: *shared fresh water resources*. This conclusion led to the development of a proposal to conduct this NATO Advanced Research Workshop.

This document has been organized into four Chapters, grouping individual manuscripts into the following sections:

1. Transboundary Water Management Problems in the Central Asian Region.
2. Management of Transboundary Water Resources in Central Asia and Caucasus Region.
3. Legal, Technical, and Institutional Aspects of Transboundary Water Management.
4. Workshop Conclusions and Recommendations.

Papers were presented and manuscripts prepared by the experts participating in Parts I, II, and III in order to provide an overview of current and emerging issues in each area. Care was taken to be sure that both the Central Asian perspective and the perspective from outside the region were presented and discussed. Efforts were also made to include participation by representatives of leading international organizations, both as technical presenters as well as participants during the discussion sessions that were organized throughout the workshop. While not achieving a perfect balance between the Central Asian and non-Central Asian perspectives, great effort was made to encourage parity and inclusiveness during this ARW. A goal of this meeting was to examine possible exchanges of ideas on how best to address these problems in Central Asia and worldwide. For it is only through clear dialogue can problems be completely understood and effective solutions recommended.

1.1. WORKSHOP DETAILS

This three-day workshop included participation by approximately 30 environmental professionals from the following countries:
- Kazakhstan
- Kyrgyzstan
- Turkmenistan
- Tajikistan
- Uzbekistan
- Azerbaijan
- United States
- Belgium
- Canada
- Italy
- Turkey
- The Netherlands

The following international organizations were also represented during the workshop:
- NATO–Security Through Science Programme
- NATO– Committee on Science for Peace and Security
- United Nations Educational, Scientific and Cultural Organization (UNESCO)
- United States Geological Survey (USGS)
- United Nations Development Program (UNDP)
- The Nature Conservancy (USA)
- The Regional Environmental Centre for Central Asia (CAREC)
- The Organization for Security and Cooperation in Europe (OSCE)
- The United States Corps of Engineers

The workshop covered a variety of topics, including the following:
- Ecologically sustainable flows for multi-jurisdictional river basins
- Shared vision modeling
- Integrated Water Resources Management (IWRM)
- Case studies of transboundary water resources management in Central Asia and throughout the world
- Legal & political bases for mediation among competing jurisdictions

1.2. BACKGROUND INFORMATION ON TRANSBOUNDARY WATER ISSUES IN CENTRAL ASIA

It is clear that the availability and distribution of fresh water represents one of the most pressing environmental and potentially social instability problems of the Central Asian region (as well as many regions throughout the world). Problems associated with sharing fresh water exist in many parts of the world and depending on the relative severity of the water misallocations, as well as the availability of possible intergovernmental solutions, these problems can induce conflicts at various levels. In Central Asia, it is believed that this phenomenon will become even more pressing as:

- Country populations grow throughout the region
- Real incomes increase (and, therefore, the demand for fresh water by households, industry and commercial users increase) as a result of oil and gas resource exploitation and other positive economic factors
- External interferences from neighboring countries confound efforts to effectively manage these regional resources (e.g., external efforts to influence the development of additional large reservoirs used for hydroelectric power generation).

The Central Asian republics can generally be classified as either "water producing" republics or "water using" republics. As a result of the relatively uneven distribution of the naturally occurring sources of fresh water (e.g., glacial snow melt in the more mountainous republics) some republics are more naturally endowed with seasonal sources of fresh water than other republics. In addition, the relatively lesser water-endowed republics are often heavily agriculturally based countries demanding large amounts of fresh water for irrigation, power generation, and other uses. Other issues pertaining to the development of new water reservoirs and associated hydroelectric generating plants make the issue of regional water allocation more complicated politically. The uneven distributions of both water supply and demand create a reasonably high likelihood for regional conflicts to occur over finite water supplies in the future. Since the collapse of the Soviet Union, Central Asia has become a tangle of unresolved transboundary water disputes. Water is considered by many as one of the most critical resources in Central Asia.

In the past, water sharing issues in Central Asia were regulated and controlled by the former Soviet system of centralized governance that set limits on water extraction and use among the Central Asian republics. In addition, and to some extent most importantly, these Soviet regulations governed the extent to which water was to be "shared" with "downstream" republics, as well as the compensation schemes that were to be used among the republics for water sharing. The rules developed by the Soviets for "water sharing" in Central Asia were based almost exclusively on production/engineering criteria that reflected the production needs for agricultural production and power generation. These engineering-based criteria, however, did not generally reflect the ecological and social benefits of the resource, leading to inefficient use and growing conflicts over the resource. Some economists argue that what is needed most to address water resource allocation problems in Central Asia are mechanisms involving "water pricing", "water markets" and other innovative multi-republic compensation schemes based on modern economic theories. More money is also needed to solve the region's water problems. As mentioned previously, in the absence of these innovative alternatives, many of these former Soviet use/consumption quotas are still adhered to by the Central Asian republics resulting in a relatively inefficient, non-sustainable and inequitable utilization of this scarce environmental resource.

An additional point of importance relates to the collapse of the Soviet system itself. With the elimination of the Soviet central government, so was eliminated the mechanism of "a higher authority" to mediate and enforce water conflicts among the republics in Central Asia. This enforcement void would represent the near equivalent of dissolution of the United States Supreme Court to rule on and enforce water conflicts among the states in the United States. Without this higher authority to address multi-republic water sharing in the region, problems have become more prevalent.

One of the world's most widely recognized examples of fresh water-related disasters is the Aral Sea situation. Misuse and inefficient uses of water for irrigation and other agricultural purposes from the Amu Darya and the Syr Darya rivers to irrigate crops in the arid areas of Central Asia caused the Aral Sea to all but dry up – shrinking some 50–75% of its volume. The failures of the Soviet central planning system can be attributed to most of the cause of this disaster, as well as other external factors. While catastrophic in nature, the Aral Sea problem is highly symptomatic of the problems facing Central Asia as the future approaches. In the absence of an effective mechanism for reconciling regional water disputes, more problems like this may occur.

One final introductory comment about the Caucasus is warranted. The neighboring region referred to as the Caucasus consists of essentially Armenia, Azerbaijan, and Georgia. The Caucasus represents a region in Eurasia bordered on the south by Iran, on the southwest by Turkey, on the west by the Black Sea, and on the east by the Caspian Sea. The Caspian Sea is the primary linkage connecting Central Asia to the Caucasus. As such, this hydrological linkage, its geographic proximity and other geopolitical linkages call for inclusion of this region in the discussions at this ARW. Beyond these physical justifications, the water sharing conflicts and experiences of these three countries bring a wealth of information to this ARW (interested readers are directed to the papers presented by Professors Campana and Suleymanov in this document).

1.3. BACKGROUND AND JUSTIFICATION FOR THIS ARW

The title of this NATO Advanced Research Workshop is "Facilitating Regional Security in Central Asia through Improved Management of Transboundary Water Basin Resources" (key words: regional, security). This workshop, supported under the NATO Committee on Science for Peace and Security, has as a goal to provide support to the 'Central Asian Nations' decision-making by providing a forum for bringing together scientists, experts, and policy makers to address critical problems associated with transboundary water management, and by creating networks of experts, and stimulating NATO-partner cooperation. The primary concern that motivated the organization of this ARW was avoidance or mitigation of regional conflict and the steps needed to attain this goal involve a more comprehensive understanding of transboundary water management issues, in general and as these issues are manifested among the Central Asian republics. The typical strengths of participants of NATO Advance Research Workshops are in the realms of technical and scientific input as most participants are affiliated with academia or research arms of ministries/agencies. It was, therefore, anticipated that the primary contributions forthcoming from this ARW would be in the forms of hydrology, water resources sciences, law, economics, conflict resolution theory, and other forms of scientific input.

2. CONCLUSION

The Directors of this ARW would like to thank the participants for their significant and important contributions to this important topic. It is recognized that water resources sharing problems are wide spread, highly complex and evolving; however events like this NATO-sponsored Advanced Research Workshop can hopefully help to address current problems and forestall the development of serious conflicts before they can germinate. Certainly, one of the more productive outcomes of this workshop is the interaction of environmental scientists from Central Asia, the Caucasus, and from NATO member countries who otherwise may not have met professionally, and therefore not had opportunities to collaborate on such an important topic as fresh water resources. The importance of these potential scientific collaborations will grow particularly as these important resources become relatively scarcer in the future, and as the need to determine effective, equitable and efficient resource sharing approaches increases. Unassailable and scientifically defensible bases for environmental solutions to problems involving multi-political jurisdictions can become a linchpin in their acceptance and implementation.

PART I: TRANSBOUNDARY WATER MANAGEMENT PROBLEMS IN THE CENTRAL ASIA REGION

PROBLEMS OF WATER RESOURCE MANAGEMENT IN CENTRAL ASIA

ERNAZAR J. MAKHMUDOV, ILHOMJON E. MAKHMUDOV, AND LENZI Z. SHERFEDINOV
Institute of Water Problems of the Academy of Sciences of Republic of Uzbekistan

Abstract: The article analyses the conditions of formation and use of water resources of the Aral Sea Basin transboundary rivers. A modern water resource management system is described. The problems related to water resource management in new conditions of development of the independent States of the Aral Sea Basin are also discussed. The issues related to ecological status of transboundary water entities and the ways on how to organize ecological management are analyzed. A new water-management strategy to improve the Aral Sea Basin water resource management is proposed.

Keywords: Water resource management; transboundary water resources; Central Asia; water flow formation

1. Introduction

Water resources, as any other natural resource, represent not only an ecological category, but an important economic category. In Central Asia, water resources acquired this significance in ancient times, about 6,000–7,000 years ago, along with the appearance of irrigated agriculture. At the beginning of the 20th century, when the industrialization process (including construction) started in Central Asia, almost half of the river flow was utilized for irrigation. The process of allocation of water resources had been completed by 1990. It is necessary to note that the process of allocation of water and energy resources of the rivers was performed in accordance with special programs (see, e.g., refs. [1–4]). The implementation of these programs resulted in utilization of entire water resources of the South area of the Aral Sea basin, and ultimately in the destruction of the Aral Sea as a morphological and ecological component of the region.

2. Water Resources Formation and Use in Economy

Prior to independence of the Central Asian states, the water management conditions in the region were characterized by data presented in Tables 1, 2, and 3.

TABLE 1. Water resources of river systems (in zones of formation, km^3/year [3,4,5,6,8])

#.	River name (river section line)	Runoff variation coefficient (CV)	Runoff rate (or water resources at 50% flow probability)	Water resources at 90% flow probability
1	Water runoff in upstream areas of the Amu Darya	0.14	65.9 (65.4)	54.4
1.1	Including Pyanj (Downstream Pyanj)	0.12	33.1(32.9)	28.1
1.2	Vahsh (Tutkaul)	0.14	19.9 (19.9)	16.4
1.3	Kunduz(Askarhona)	0.23	3.50 (3.42)	2.50
1.4	Kafirnigan	0.19	5.51 (5.41)	4.19
1.5	Surkhandarya	0.19	3.67 (3.60)	2.80
1.6	Sherabad	0.32	0.22 (0.22)	0.14
2	Runoff in the rivers situated in the South-West of Uzbekistan		6.55 (6.48)	5.17
2.1	Including Zeravshan	0.15	5.28 (5.24)	4.30
2.2	Kashkadarya	0.26	1.27 (1.24)	0.87
3	River runoff in Turkmenistan		2.77 (2.60)	1.35
3.1	Including Murgab	0.29	1.54 (1.50)	1.0
3.2	Tedjen	0.58	0.96 (0.86)	0.35

(Continued)

	Total in the Large Amu Darya Basin (excluding North Afghanistan rivers which do not flow in)	75.2 (74.6)	60.9
4	Water runoff in upstream areas of the Syr Darya	(25.56)	19.59
4.1	Including Naryn (Uchkurgan)	(13.74)	9.9
4.2	Karadarya (Kampyravat)	(3.76)	2.24
4.3	Inflows in Fergana Valley	(8.06)	7.45
4.4	Middle part rivers runoff	(9.3)	6.51
4.4.1	Runoff from Golodnaya Steppe foothills	(0.6)	0.31
4.4.2	Chakir River runoff	(8.7)	6.2
4.4.3	Including Chirchik (Khodjikent)	(7.1)	4.8
5	Water resources in the area of formation in relation to the section line of Chardarya (excluding runoff from Golodnaya Steppe foothills rivers)	(34.26)	25.8
6	Arys and rivers of Karatau mountain ridge	(2.3)	1.5
7	**Total in the Large Syr Darya Basin**	(37.2)	27.6
8	**Total in the Large Amu Darya and Syr Darya basins**	(112.4)	88.5
9	**Total in the large Amu Darya and Syr Darya rivers flowing into the Aral Sea**	(100.11)	80.2

TABLE 2. Main indicators of water balance in the Amu Darya Basin (area in thousand hectares, water volume in cubic meters per year) [4,5,6]

#.	Indicators	Level in [x] 1990г. According to "scheme ....." (4. p.233)	Estimates of actual values in 1990 – year ~50% [x] probability of water availability	Deviations (±)
1	Available water resources (2/2.1+3+4)	72.5	74.5	+2.0
2	River water resources	54.4	65.8	
2.1	Including minimum flow	62.1		
3	Ground water resources (output)	1.5		
4	Return water resources	8.9	8.7	−0.2
5	River water intake for irrigation	50.2	57.4	+7.2
6	Internal use of return water for irrigation	10.5	2.7	−7.8
7	Water intake for industrial and residential use	10.4	2.7	−7.7
8	Discharge of return waters into local water intake units	9.8	9.3	−0.5
9	Total water losses in river beds and water reservoirs	8.7	2.9	−5.8
10	Water flushed into the Aral Sea through river beds	3.2	6.2	−3.0
11	Irretrievable losses and withdrawals (5+7+9)	69.3	65.7	−8.5
11.1	Including irrigation agriculture (5−4)	41.3	62.4	+21.1
12	Area of irrigated land	3514	3430	−134
12.1	Irrigation norm – gross, m³/hectare per year, ((5+6):12)	~17270	~17520	~+250
12.2	Specific irretrievable losses, m³/hectare per year, ((5−4):12) [+] excluding the Zeravshan River and Kashkadarya River basins	~11750	~14200	~+2450

TABLE 3. Main indicators of water balance in the Syr Darya Basin (area in thousand hectares, water volume in cubic meters per year) [3,5,6]

#	Indicators	Level in 1990r. According to "scheme ....." (3. p.233)	Estimates of actual values in 1990 – year ~50% probability of water availability	Deviations (±)
1	Available water resources (2/2.1+3+4)	48.1	49.8	−1.7
2	River water resources (90% probability of water availability)	27.6	34.3	−10.6
2.1	Including minimum flow	33.0	33.0	−
3	Ground water resources (output)	3.1	H.c	−
4	Return water resources	12.0	15.5	+3.5
5	River water intake for irrigation	36.6	40.9	+4.3
6	Internal use of return water for irrigation	6.5	6.6	+0.1
7	Water intake for industrial and residential use	2.1	2.4	+0.3
8	Discharge of return waters into local water intake units	0.6	3.3	+2.7
9	Total water losses in river beds and water reservoirs	3.8	4.0	+0.2
10	Water flushed into the Aral Sea through river beds	2.9	2.0	−0.9
11	Irretrievable losses and withdrawal s (5+7+9)	42.5	47.3	+4.8
11.1	Including irrigation agriculture (5−4)	24.6	25.4	+0.8
12	Area of irrigated land	3094	2983	−91
12.1	Irrigation norm – gross, m^3/hectare per year, ((5+6):12)	13930	15923	+1993
12.2	Specific irretrievable losses, m^3/hectare per year, ((5−4):12)	7950	8515	−565

3. Management of Water Resources Consumption

After the Second World War, social and economic development in the region was directed toward a complete use of the irrigating capacity of the Amu Darya River (~4.3 million hectares) and Syr Darya River (~3.0 million hectares). The river flow management was designed mainly for satisfying irrigation needs. By the 1990th, the level of diversion of water from Syr Darya River and Naryn River reached 92–94% of the "norm" by providing the guaranteed diversion of 32–33 km^3 per year during 90 years. The "norm" for the Syr Darya is estimated at 37±2 km^3/year. For the Amu Darya River, the flow diversion values were slightly lower (80–85%) of the "norm" of 75±4 km^3/year, and the guaranteed annual level of water diversion was 90%, i.e., less than 60–64 km^3/year.

The hydraulic facilities were also designed for the use of hydroelectric potential. The hydroelectric potential in the region, at an achieved level of technological development is estimated as 580–590 TW/hour. The major part of this (over 50%) belongs to Tajikistan and Afghanistan (over 50%); 25% to Kyrgyzstan; 15% to Uzbekistan; 4% to Turkmenistan; and 3% to South Kazakhstan [1,2]. The economically accessible part of the hydroelectric potential equates to about a quarter of its value, and it is only about 6% [1] of the hydroelectric has been used so far. In the long term, taking into account expected increases in the cost on energy resources, the economically accessible part of the hydroelectric potential will become equal to the technically utilized portion, and this will amount to 50% of its value. Taking into account the current prices for electric power, its annual output can be estimated as $5–10 billion. However, this potential still needs to be developed at approximate cost of $800–1,000 per 1 kW. The use of hydroelectric potential will allow the whole region to satisfy the major part of its needs in electric power demand using renewable resources. However, it should be noted that presently the contribution of the irrigated agriculture into the gross domestic product of the region is significantly higher than the contribution expected from hydraulic power development.

It should be stressed that the development of water resources of the region was organized and conducted in accordance with special programs of complex development and protection of water resources.

The results of implementation of these programs in the Amu Darya and Syr Darya rivers are shown in Table 2 and Table 3.

To manage the river flow, a system of water reservoirs was constructed in the region (Table 4). Allocation and consumption of water in large river basins are shown in Tables 5 and 6.

For the purposes of irrigation, hydraulic power development, and other branches of economy, the river systems of the region were transformed into hydro-economic systems, whose main assets are still estimated in the billions of dollars. This development of hydro-economic systems was performed in accordance with a plan, and the ultimate result of this program was the planned inter-zonal diversion of river runoff (particularly, the diversion of a part of Siberian rivers runoff to Central Asia).

4. Ecological Condition of Waterways

Today, the overall ecological consequences of water resources development in the region are assessed as negative and their main result is seen in the Aral Sea destruction. However, the economic conesquences include a 2–3 time increase of the irrigated land and the hydroelectric resources development. It is evident, that in view of the demographic growth in the region, the social and economic consequences of the irrigation should not be assessed as completely negative.

It is better to refine everything once again and "separate the wheat from the chaff". But it is necessary to note an accuracy of the hydro-economic programs of that period: everything that happened, including the expected situation, time, initial scope, and other parameters of the runoff diversion, had been calculated beforehand practically with no mistakes. But the main shortcoming of this region's economy – high, unacceptable for arid zone expenditure of water – has not yet been overcome.

The formation of a state independence of Uzbekistan started in very complicated social, economic, and ecological conditions. The condition of the water industry of the country, as well as the condition of all water consuming branches of economy, and the condition of its infrastructure was assessed as problematic and even critical.

Uzbekistan, as a part of Central Asia and one of the most arid regions of the world, by the time of gaining its independence used about 52–57% of its very limited water resources which were almost fully used in the economy. The existing significant shortage of water resources was aggravated with their exhaustion in water quality. This process is quite complicated, and it is accompanied by salinization and contamination of surface and ground waters. The most apparent process in the region is the process of water salinization that, according to its genesis, depends on the evaporative concentration. The evaporative concentration conjugating with mass exchange of the first type results in increase of mineralization of natural waters. In the Syr Darya River Basin these phenomena were observed during the 20th century, and by the end of the last century, the average annual mineralization in the Syr Darya River at the Kayrakum reservoir outlet point was equal to 1.22±1.3 g/l, and the mean waters were practically unsuitable for drinking water supply.

In the Chirchik, water salinization processes practically reached the longitude of the city of Yangiyul where a mean water mineralization reached all allowable limits, and in the mouth it exceeded these limits In the Zeravshan, water salinization processes practically reached the river section line of the city of Khatyrchi, and rapidly approaching the Damkhodjinsk section line. The main water intake facilities of Navoi and Bukhara Oblast are now located downstream of this line.

For almost 50 years in the Kashkadarya river bed, the brackish tail water of Chimkurgan reservoir has been flowing. The same situation is arising in the downstream part of the Surkhandarya River.

In the downstream part of the Amu Darya River, in the section line of the Tuyamuyun reservoir the average annual mineralization of river water almost reached allowable limits, and in the mean waters it exceeded them. The mineralization of the Amu Darya River waters at the upstream outlet point is also increasing. However, it has not reached the maximum allowable limits.

The mineralization is increasing due to growing concentration of magnesium ions, sodium sulfates, and sodium chlorides. According to these indicators and, also, to the total hardness, the water becomes unsuitable for drinking water supply and very often, for irrigation too. Increased incidence rate of cholelithic and nephrolithiasis diseases to a considerable degree is related to an increase of total hardness of drinking water.

On the territory of Uzbekistan, there are various sources of natural water contamination including industry, power engineering, motor transport, and other engineering services, mining and processing of

TABLE 4. Technical indicators of complex hydro schemes of the Aral Sea Basin River systems [1,2,6]

№ п/п	Basins, water reservoirs	Water resources in relation to river section line (km³) Runoff rate	Water resources in relation to river section line (km³) 90% probability of water availability	Reservoir's characteristics (km³) Full capacity	Reservoir's characteristics (km³) Usable capacity	Hydroelectric Power Station Installed power in megawatts.	Hydroelectric Power Station Produced electric power in thousand W/Hour
1	2	3	4	5	6	7	8
1.	Syr Darya River Basin [x]			35.5	27.3	2168	8.0
1.1	Toktogul water reservoir	13.7	7.9	19.4	14.0	1200	4.0
1.2	Andijan water reservoir	3.76	2.24	1.75	1.6	140	0.6
1.3	Kayrakum water reservoir	25.5	19.6	3.4	2.5	126	0.5
1.4	Charvak water reservoir	7.10	4.80	2.0	1.6	600	2.5
1.5	Chardarya water reservoir	34.9	26.1	5.5	4.70	100	0.40
1.6	Water reservoirs located on small rivers (tributaries)			3.30	2.90		
2.	Amu Darya River Basin [x]			19.4[xx]	11.02	2850	11.9
2.1	Nurek water reservoir (1975)	19.9	16.4	10.5	4.5	2700	11.2
2.2	Tuyamuyun water reservoir (1985)	69.2	57.0	7.3	5.1	150	0.7
2.3	Water reservoirs located on small rivers (tributaries)			1.6	1.42		
2.4	Rogun water reservoir (under construction)	19.0	16.0	11.8	8.6	3600	12.5
3.	Total in the Aral Sea Basin [x]			66.7	46.9	8618	32.4
3.1.	Including working reservoirs			54.9	38.3	5018	19.9

[x] Excluding zero-discharge rivers (Arys, Zeravshan, Kashkadarya, and rivers of Turkmenistan)
[xx] Excluding Rogunsriy water reservoir

minerals and hydrocarbon materials, farming industry, processing of products of plant cultivation and livestock farming, and housing and communal infrastructure related to "utilization" of various domestic wastes. Because of that, on the territory of Uzbekistan all types of natural water contamination are present: biological, chemical, radioactive, thermal, and their complex combinations.

TABLE 5. Indicators of transformation and use of water resources distributed among hydro-economic regions of the Amu Darya River basins in the years of ~50% and 90% probability of water availability [6]

#	Indicators	Water volume, km^3/year	
		50% probability of water availability	90% probability of water availability
1	2	3	4
	I		
1.	Available upstream water resources	71.9	60.5
1.1	River water inflow into the HER (hydro-economic region)	65.8	54.4
1.2	Return water resources (formed within boundaries of HER – flushed into rivers)	6.1	6.1
2.	Irretrievable losses	8.3	8.3
3.	Runoff from HER	63.6	52.2
	II		
1.	Available midstream water resources (excluding the Zeravshan, Kashkadarya rivers, etc.)	65.8	54.3
1.1	River water inflow into the HER	63.6	52.2
1.2.	Return water resources (formed within boundaries of HER or flushed into rivers)	2.2	2.1
1.2.1.	Including the amount flushed from Karshi and Bukhara oasis	1.1	–
2.	Irretrievable losses, including water taken by Karakum, Bukhara and Karakul channels	32.3	30.3
2.1	Including water taken by trunk channels	26.2	26.0
3.	Runoff from HER/Losses along the river bed and in Tuyamuyun reservoir	31.6/1.9	24/1.9
	III		
1.	Available downstream water resources (Khorezm, Karakalpakstan, Dashtkhauz)	32	24.4
1.1	River water inflow into the HER	31.6	24.0
1.2.	Return water resources (formed within boundaries of HER or flushed into rivers)	0.4	0.4
2.	Irretrievable losses	26.1	20.9
3.	Runoff from HER (to the sea or local reservoirs)	5.9	3.5
3.1.	Actual runoff in Kysyljar River section (1990.)	6.12	

TABLE 6. Indicators of transformation and use of water resources distributed among hydro-economic regions of the Syr Darya River basins in the years of ~50% and 90% probability of water availability, [6]

#	Indicators	Water volume, km³/year	
		50% probability of water availability	90% probability of water availability
1	2	3	4
	I		
1.	Available upstream water resources (Fergana HER and the Naryn and Karadarya rivers)	32.8	26.6
1.1	River water inflow into the HER	25.5	19.6
1.2	Return water resources (formed within boundaries of HER or flushed into rivers)	7.3	7
2.	Irretrievable losses	9.2	9.0
1	2	3	4
3.	Runoff from HER	16.3	12.6
	II		
1.	Available midstream water resources, including CHAKIR	28.2	20.5
1.1	River water inflow into the HER	24.5	18.0
1.2.	Return water resources (formed within boundaries of HER or flushed into rivers)	3.7	2.5
2.	Irretrievable losses	13.2	12.0
3.	Runoff from HER	11.3	8.5
	III		
1.	Available downstream water resources (excluding Arys)	11.8	9.0
1.1	River water inflow into the HER (Chardarya reservoir section)	11.3	8.5
1.2.	Return water resources (formed within boundaries of HER or flushed into rivers)	0.5	0.5
2.	Irretrievable losses	9.3	~8
3.	Runoff from HER (Kazalinsk River section)	3.5	-

Bacterial contamination is frequently connected with pathogenic germs and viruses. Today's situation of this kind of contamination is elevated and is characterized by the fact that almost all surface water supply sources require sterilization. The state of the ground water sources is also becoming a more complicated issue.

The chemical pollution becomes apparent much more in the water bodies of Uzbekistan from industrial, agricultural ones, and municipal sources. The water is contaminated with heavy metals, cyanides, thiocyanates, metabolites of pesticides, herbicides, and other ingredients the concentration of which is regulated by the general, sanitary, and organoleptic norms. There is an increase in radioactive water contamination, e.g., there is increase of uranium concentration in the Aral Sea area where it reaches $10^4 \div 10^{-5}$ g/l.

Thermal water pollution results from the thermoelectric power stations with the direct cooling system and metal manufacturing premises.

Such effects as the Aral Sea degradation and degeneration of its ichthyofauna, changes of specific phyto- and animal plankton structure in downstream areas of large and average rivers of Amudarya, Surdaya, Zerafshan, Chirchik, and others are the results of complex combinations of depleting of water resources both in quantity and quality. For example, the reduction of fish that does not meet veterinary–sanitary requirements usually takes place in the return water lake-stores systems.

In general, it is necessary to take into account that a progressing pollutant concentration is accompanied by cumulative effects resulted in degeneracy of important biocenosis components.

5. Potable Water Supply Problem

A quite complicated situation regarding human water supplies has been formed in this hydro-chemical background.

The problem of good quality potable water is one of the most pressing ecological issues that needs to be solved by Uzbekistan and other states of the region.

In Uzbekistan, only 75% of cities and urban and rural communities have water supply systems. Only 85% of urban population uses a centralized water supply system and only 60% in the rural areas. An average daily water consumption per urban resident in 1990 was 470 l/d, including 750 l/d in Tashkent and 388 l/d in other cities and urban and rural communities.

It can be shown that in Uzbekistan the use of water for domestic purposes is 1.9–2.0 km^3 per year. However, the issues related to high-quality potable water need to be solved in all regions, first of all, in the Aral Sea area, and in the lower areas of the Zeravshan (Navoi and Bukhara oblasts) River and the Kashkadarya (Kashkadarya oblast) River, where not only groundwater, but also surface water are often of low quality and used for the agricultural-domestic purposes.

6. Water Management Strategy

It is evident that there is a need to develop a water management strategy for the whole region and the states both included in the region and bordering with that region. Such a strategy should be a rational system of high priority (incorporating political, defense-strategical, economical, and sanitary-hygienic components) purposes and technologies of prospective (and operative) water supply management of the countries (region), organization and control of water-consumption and water-use.

The strategy should also take into account natural, social-cultural, economical and technical-technological capacities and geopolitical conditions.

In the past, the strategy of water management activities, being an empirical generalization, was based on the social-cultural traditions of the region. Later during the soviet period the strategy was formed on the bases of various kinds of legislative and other normative acts. Nowadays, in compliance with provisions of the current law, the strategy is to be determined on a state (and if necessary, on a national) level of management. The strategy defines the long-term social approaches and requirements to water as an essential life-supporting natural resource and as the habitat of many biological hydro coles species, and at the same time, as a potentially dangerous and harmful factor due to its nature. Moreover, in arid environments, the water management strategy functionally interconnects the structures of the economic and ecological security of the country.

The strategy is implemented (and simultaneously revised, corrected, etc.) through long-term planning of the water management activity in specific programs. The water management planning includes the theory and practical issues of water supply of the country (region); transition to optimal water-use and water-consumption, taking into account the regularities of forming and peculiarities of using of water resources; special measures for providing water supplying, water-use and water-consumption and preventing dangerous and harmful water related effects to public and environment. In addition, it is necessary to analyze the interconnections of the water management systems and complexes, and to forecast their technical-economical parameters; to formulate the requirements to the systems and services providing monitoring and control and to determine a structure and sequence of their activeties. The water resources are to be distributed to the territorial-production complexes and water management complexes of the country (region) for social, ecological, sanitary needs. The protection of water against depletion, pollution and their ecological well-being rehabilitation is planned.

The main feature of contemporary water management is seen in its ecologiozation and in inclusion of economic mechanisms and legal regulations in the program implementation. In general, the water management program presents a new scientific and technical foundation and proposes solutions to problems of rational using and water conservation, and all spectra of issues related to effective use and conservation of the environment. The strategy realizes the provisions of water management doctrines and strategies developed by governmental and legislative bodies. If necessary, recommendations on improving of water relations are to be developed to take into account new data on water needs; availability of new technologies; water-use and water-consumption on the one hand, and on the other hand– ecological limits and imperatives. The water management program is to be reviewed and approved by the government of each country.

It is obvious that the water management programs shall proceed from the fact that the social-economic perspectives of the region's members are to be based on the regulations tested by the whole world community:

- Sustainable increases in the common wealth can be achieved by implementing power and resources saving technologies and approaches, and minimizing of the production and service costs
- Restructuring of the countries' economies is directed on absorption of their own labor resources and development of a high technology manufactures, forming of new industries based on new water saving technologies, or waterless technologies, low-waste and no-waste technologies
- Ecologization of all spheres of production of material goods and vital functions of the society for conservation of biota diversity, the natural vector of ecotopic and biocoenotic selection, and as a whole, the hygiene and quality of the entire environment.

In this situation, Uzbekistan should take into consideration three factors that are connected to the formation of quantity and quality of water in the intergovernmental relations on development of the water resources management strategy.

1. The natural river flow is formed out of Uzbekistan, mainly in the chain of mountains of the Tien Shan and the Pamir.
2. The main water reservoirs that transform the intrinsic regime of the river flow for household use needs and providing the water supplying stability are located outside of Uzbekistan.
3. Original water-management systems were designed to service water-management complexes for irrigation and power purposes, and the main regime of their functioning was to meet first the needs of irrigated farming at the expense of decreasing of power production. This is not profitable to the present owners of these hydro-facilities.

In the former Soviet Union, the mentioned above issues and following from them interrelations between the Aral Sea area states were centrally regulated. Nowadays, when Uzbekistan became an independent state and international proprietor, there is an urgent need to regulate in the water relations regulation between the Aral Sea Basin states by the highest level international/legal authority. It is evident that it is in the national interests of all countries of the basin to define the regulation of the water-management systems functioning that could take into account the mutual economic and ecologic needs and benefits, and damages.

According to some forecasts, the issue of water supply of Uzbekistan could be aggravated, especially in the long-term perspective, of 2025–2030 due to global climate change. However there also some "bright" forecasts. According to forecasting scenarios, the river flow changes could be reduced one third from its standard (about 32–38 m^3/year) level. This would reduce sharply the irrigation capacity of the water management systems, and, to a greater extent, taking into account the territorial location of Uzbekistan at the zone of transit and dissipation of the flow; it would have a strong affect on the water supply.

The current almost exhausted water resources and the possibility of negative changes in the river systems of the region require a radical change of the established approaches to water-use and water-consumption and overcoming of the Aral ecologic disaster.

Under these circumstances Uzbekistan should focus its strategy on saving the water economy in the all branches of the economy and especially in the irrigated agriculture that consumes more than 90% of water resources.

The rehabilitation of irrigative water presents an important element of water saving. High quality of irrigative water improves the harvest of cultivated crops, reduces the expenses for overflowing of used water from irrigated fields, affects positively on the adjacent natural and cultivated landscapes. However the issue of water quality management of the irrigation sources of large rivers does not have an ultimate scientific/objective solution. There are some approaches to river flow demineralization that are used in practice, but they can solve only special problems, and, very often, only temporarily. In Uzbekistan, the solution to the river flow demineralization issue is estimated to be equivalent to adding into the fund of irrigative farming from 200,000 to 500,000 ha.

A decisive component of water supply management is a reliable water supply. The demineralization of river water would greatly increase the water supply effectiveness. In addition to demineralization, in some regions of Uzbekistan the growing ecologic tension requires to use additional reprocessing facilities for water desalination and purification in order to meet the sanitary standards for water quality.

According to some forecasts, it is expected a decrease of the water flow in the region. In this situation it becomes necessary to utilize the groundwater. The way of current ground water use is difficult to characterize as a rational.

The shortage of water resources was the main cause of the Aral Sea problem. In view of an expected decrease of water flow in the region, the conservation and further restoration of the Aral Sea may be an unachievable objective. At the same time, there is a self consistent solution of the Aral Sea problem. Within the natural structure of Central Asia the Aral Sea used to perform the functions of the largest regional salt receiver. First of all it is necessary to restore this function of the Aral Sea. To do this it is necessary to stop salination of natural and artificial waterways and reservoirs in the deltas of the Syr Darya and Amu Darya. This should significantly improve the ecologic situation in the Aral Sea area and, along with activities aimed at demineralization of river waters, would positively affect on the productivity of irrigated lands in Karakalpakstan and Khorezm. The quality of irrigation water is becoming the determining factor of the environment rehabilitation.

7. Conclusion

The current and future hydro-economic and hydro-ecologic conditions in Uzbekistan are closely related to each other and very important to life-support of the population. It demands the development of the new water management program for Uzbekistan which is based on achievements of the world community.

Realization of strategic water management programs, as is known, occurs not in 1–2 years, but will require decades. Therefore it is needed already now to target scientific and design/ research teams for development of mid- and long-term programs which could presented for approval by statesmen and community. Well-being of the community of the region, its social-economic and ecological security depends on the timely development of these programs and appropriate strategic decisions and measures taken by the states. It is out of question that the development and implementation of the water management program should be performed by each state of this region separately. Only joint intellecttual and technological efforts and sharing of natural resources can justify the public expectations of prosperity, well-being and peace. This approach is justified by past experience and new achievements in the field of water resource management.

References

Aral Sea Basin Hydropower, Press-release. Tashkent: Tashgidroproject, 1994, p. 28.

G.I. Kornakov, S.A. Borovets, A.A. Bostandjoglyu, R.I. Bakhtiyarov. Existing state and perspectives of development of main branches of the national economy in the Amu Darya River Basin, Tashkent: Saohydroproject, 1968, p. 114.

Scheme of complex use of water resources of the Syr Darya River Basin, Tashkent: Sreadazgiprovodhlopok, 1969, p. 126.

Refinement of Scheme of multipurpose utilization of water resources of the Amu Darya River Basin, Tashkent: Sreadazgiprovodhlopok, 1983, p. 372.

N.R. Khamrayev, L.Z. Sherfedinov: Central Asia Water Resources: assessments, scale of development, variability, significance in respect to ecological security and social and economic development of Uzbekistan, In the book: "Water problems of arid territories", issue No.2 (Institute of Water Problems) Tashkent: "Uzbekgidrogeologiya", 1994, pp. 3–18.

L.Z. Sherfedinov, N.G. Davranova. Water is a limiting strategic resource of social-economic and ecological security of Uzbekistan, In book: "Water reservoirs, force majeure and stability issues, Tashkent: University, 2004, pp. 123–133.

COMMUNICATING THE ISSUES OF THE ARAL SEA BASIN
LONG-TERM VISION FOR THE ARAL SEA BASIN
THE ARAL SEA BASIN MANAGEMENT MODEL (ASB-MM)

JOOP DE SCHUTTER
UNESCO-IHE Institute for Water Education
Head Department of Water Engineering
The Netherlands

Abstract: The collapse of the Soviet Union and the subsequent political and economic separation of the countries of the Aral Sea (i.e., the Central Asian republics) have changed the situation with respect to shared management of the basin water resources dramatically. The countries were very quick to identify and develop new political structures based on their own national identities and entrance into the global arena, while at the same time they were under extreme pressure to avoid economic collapse. These changes and developments led to a situation of transition from existing agreements and operational policies for water sharing in the Aral Sea Basin towards a situation based on the diverging interests of the individual states and their main water users. This situation became quickly recognized by many scientists and water and irrigation works managers in the region, who started to report on the dangers of this new situation and the short term negative impacts this would have to the, still mainly irrigated agriculture based, economies of the countries in the region.

Keywords: Aral Sea Basin, basin-level management and modeling, water management capacity development

1. Water Resources

The prosperity and growth of the Countries of the Aral Sea Basin in Central Asia depend primarily on the stability of the areas of formation of its main rivers in the ecosystems of the Pamirs, Tien Shan, and Altai. These mountain ecosystems absorb moisture and act as gigantic freshwater accumulators. However, ecological degradation and climate change are increasingly causing constraints to the stability of the water resources and ecosystems in the upper watersheds and in the Aral Sea Basin as a whole. The glaciers of the Pamirs and Altai lost over 25% of their ice reserves between 1957 and 2000, which will cause river discharges to increase in the short term, but decrease over the years while presently already very low precipitation figures will become even lower. The river discharges of the Amudarya and of the Syrdarya have been regularly monitored since 1914 as shown in the figure below. Extremes have occurred between 1972–1991 (dry) and 1952–1971 (wet), but also the regular discharges throughout the years show very large volatility and make water users vulnerable.

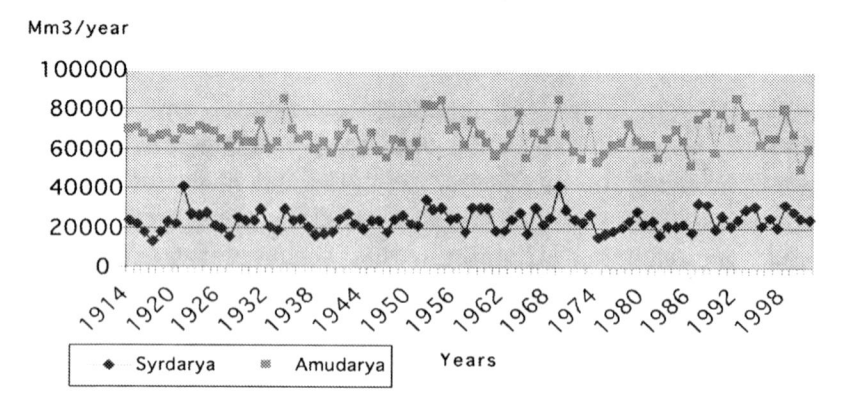

Figure 1. River discharges of the Amudaya and Syrdarya over the period 1914–2001. (After REBASOWS 2006.) (6)

Irrigated agriculture in Central Asia has a history of more than five thousand years. However, during the last forty years almost all the water that used to come down from the mountains and went to the Aral Sea is used for intensified irrigation and this has resulted in the widely known Aral Sea crisis. The decreasing amount of water from the rivers has started to affect the continuity and level of operation of irrigated agriculture in the region with increasingly serious consequences for the socioeconomic conditions of water users and for the regional economies as a whole. Equally the ecosystems of Central Asia are also suffering and the conflicting interests of water use for irrigation and hydropower have become more prominent. The main issues at stake are how to moderate water demand for irrigation on the one hand and how to find a balance between irrigation and hydropower water release requirements on the other, with water demand in support of other functions at more distance even. Consequently, drinking-water quality, ecosystems, soil fertility and crop yields are deteriorating, while poverty, unemployment and migration are on the increase in the rural areas. Typical examples of this process can be found in the deltas of the rivers North (Kazalinsk, Aralsk) and South (Karakalpakstan) of the Aral Sea, where living circumstances have become extremely difficult at times.

2. Recent History

The high demand for water for irrigation is not the only problem that the region is facing and many other issues make finding sustainable solutions very challenging. When looking at the events that have led to the breakup of the Soviet Union in 1991 the region had already known the rise and fall of many different states such as Ariana, Baktria, Sogdiana, Bokhara, Khorezm, and many others (2) that saw no conflicts over water use even up to the colonization of Turkestan by Tsarist Russia. This situation only really changed during the last 40 years of Soviet rule that started managing water resources on a centralized basis while at the same time developing new and large scale infrastructure, without paying adequate notice to financial feasibility and physical sustainability of the systems it introduced. However, these new systems required the development of a very strong scientific basis for irrigation and integrated water resources management in the region, which is still present and functional. In the beginning of the nineteen eighties the Aral Sea crises became apparent. Governments in Central Asia started to work on plans to deal with the situation which led to the approval of the "State Program on Priaralye" in 1986. This plan led to the installation of the river basin management organizations (BVO) and started to allocate financial means to research and projects aimed at mitigation and improvement of the situation.

Of course the collapse of the Soviet Union and the consequently political and economic separation of the countries of the Aral Sea have changed the situation with respect to shared management of the basin water resources dramatically. The countries were very quickly to identify and develop new political structures based on their own national identities and entrance into the global arena; new economic policies based on for example more or less liberalization and own sector priorities (e.g., agriculture, industry, mining, and hydropower), while at the same time they were under extreme pressure to avoid economic collapse. All these developments led to a situation of transition from existing agreements and operational policies for water sharing in the Aral Sea Basin towards a situation based on the diverging interests of the individual states and their main water users. This situation became quickly recognized by many scientists and water and irrigation works managers in the region, who started to report on the dangers of this new situation and the short term negative impacts this would have to the, still mainly irrigated agriculture based, economies of the countries in the region.

As a result of this regional consultative process and stimulated by international interests the five Central Asian States decided to start to join forces to address the huge challenges that they face to achieve sustainable development on the basis of sound water management and many initiatives have been developed over the last decade. The main institutional vehicle for these initiatives became the Interstate Commission for Water Coordination (ICWC) that was established on the basis of an "agreement for collaboration on joint interstate water resources management in 1992". In the same period, the five states confirmed their agreement that water allocation was to be based on the existing water resources in accordance with equity and mutual benefit. ICWC in principle became the collective body responsible for water allocation among countries on the basis of monitoring, research and assessment of water allocation proposals (technical, financial, ecological, legal, and institutional) mutually agreed among states.

The ICWC is supported by a Scientific Information Center (SIC) that works closely together with the main river basin management organizations (BVO) for the Syrdarya and Amudarya rivers, with the relevant ministries of the partner countries and other partner organizations in ICWC. The SIC has branch offices in all partner countries. In addition the SIC produces and shares information (website, bulletins, journals, training centre). It is responsible for preparation and organization of all the meetings of the commission and coordination of implementation of decisions. Since 1997 the Executive Committee of the International Fund for the Aral Sea (IFAS) is responsible for implementation of the Aral Sea Program, which is a cluster of project initiatives commonly implemented in the basin. The responsibilities of all interstate organizations in the Aral Sea Basin have been agreed to during a heads of state conference in Ashgabat in 1999.

At the same time existing documents are not sufficient to guarantee adequate water allocation, use and control, because the present agreements and laws do not take into account all the issues at stake. There are for example no regulations for water allocated to the Aral Sea and issues of water use inefficiencies and emergencies have not been identified nor have they been settled. The water policies for example of Turkmenistan and Uzbekistan with a very prominent role for state regulations differ completely from those in Kyrgyzstan and Kazakhstan. Discrepancies between the national water laws cause many problems between the countries with the extreme example of a water export law in Kyrgyzstan that requires states to pay for water that they import from this country.

3. Pressure Indicators

Meanwhile the overall development in Central Asia translates into strong socioeconomic growth figures for most countries of the region that in turn will affect future water use. The base development indicators including land use changes, degrading ecosystem dynamics and population growth all indicate increasing water stress and hence the need for changing water allocation and water management policies. The following table shows the historic parameters for land and water resources development in the Aral Sea Basin since 1940.

It can be concluded that without proper counter measures the situation with respect to water use in Central Asia will quickly aggravate, which may in turn cause a direct threat even to the food security situation in the region. In order to be able to start dealing with these problems political and institutional constraints may prove to be major obstacles as the usually considered technical ones. An integrated, process based, approach based on what has been achieved so far will

Indicator	Unit	1940	1960	1970	1980	1990	2000	2003
Population	Million	10.6	14.1	20.0	26.8	33.6	41.5	43,78
Irrigated area	Thousand ha	3.8	4510	5150	6920	7600	7890	7900
Total water withdrawal	km^3/year	52.3	60.61	94.56	120.69	116.27	100.87	118
Including irrigation	km^3/year	48.6	56.15	86.84	106.79	106.4	90.3	109,56
Specific water intake per 1 ha of irrigated land	m^3/ha	12800	12450	16860	15430	14000	11445	13868
Specific water intake per capita per year	m^3 per capita per year	5000	4270	4730	4500	3460	2530	2695
GDP (region)	Million US$	12,2	16,1	32,4	48,1	74,0	54,0	34.4

Water Use Dynamics in the Aral Sea Basin since 1940 (excluding Afghanistan). (After Dukhovny V.A. and Pereira L.S., 2004.)

be necessary. The main indicators that will decide future decision scenarios and strategies for the region are estimated to be:

- *Population Growth*: At current growth rates the population will double over the coming 25 years, which will place a high demand on both food (security) and water resources for irrigation as well as increased irrigation efficiencies.

- *Climate Change*: In the short term snow melting from the Himalayas will increase, but in the longer term these sources will decrease and less water will become available per year. Moreover the characteristics of the continental climate in the region show a tendency to become more extreme.

- *Hydropower*: Demand for electricity is highest during wintertime when the reservoirs of the region need to be full. However irrigation during the growth season conflicts with the necessary filling regime for hydropower. A second threat is the increasing involvement of international (U.S.A, Russia, China, and Iran) players in the hydropower market in the upper water sheds of the Aral Sea Basin that makes decision making ever more complex.

- *Natural and Man-made Disasters*: The Aral Sea Basin is an area prone to both natural (earthquakes, severe draughts) and man-made (terrorist acts, regional wars) threats that can paralyze crucial water infrastructure. In these cases large groups of the population may become victims of either direct food shortages or economic disaster.

- *Common Framework for International Water Laws*: Agreement on water sharing needs a shared vision on trans-boundary water issues and a common legal framework. The regular conventions (UN Convention of 1997, Helsinki Rules of 1966) do not seem suitable for Central Asia. As a result conflicts over water use may increase instead of being solved and may even cause territorial disputes.

- *Afghanistan*: Until now Afghanistan has not been included in the development of a common structure for trans-boundary water sharing in Central Asia, although advanced plans exist for large scale development of new irrigated land areas that may claim between 9.5 > 13.4 km^3/year.

It is clear that there is a need to find ways to diminish the amount of water used, but at the same time there is a need to offer opportunities for sustained economic growth, while improving the sociopolitical situation. This is a major challenge that can not be met by simple measures. Furthermore, interventions not only influence their targeted objectives, but usually generate important side effects that may adversely affect stakeholders in the process. In order to be able to develop sustainable solutions on the basis of agreed trade-offs it is necessary to test and communicate them before actually implementing them. In order to support such a planning and decision making process on a river basin scale high level, integrated, decision support concepts and tools are necessary. The need is for both advanced communication and planning process approaches as well as for advanced integrated models that can combine hydrology and water management options knowledge with socioeconomic development under different future development scenarios.

"The key to an approach to deal with the water management problems of Central Asia is in the combined use of scientific research and a thoroughly designed communication and planning process that will take the issues to all levels of decision making from the presidents and water ministers of the countries involved to the irrigation farmers and the population at large".

4. Central Asia Common Water Resources Vision Process

As demonstrated above and despite the huge efforts made in the recent past the need for continued and intensified evaluation and communication of integrated water management issues in the Aral Sea Basin is still urgent. The key to success is to facilitate planning and decision making at the top political level in the Aral Sea Basin countries. This will require both sophisticated and detailed technical assessments and transparent communication of the results of the technical calculations in order to maintain credibility with the target audience. At this moment in time (trans-boundary) water management can not be left to the water managers alone, but it must as well be based on commonly understood and agreed approaches that are shared with a wide range of stakeholders.

The process needs to start with the development of a common, basin wide shared long-term vision on the principles of water resources used in the Aral Sea Basin. Within the wider framework of this long-term vision shared development policies can be developed that fit the overall picture for the future. At the same time there is a need to develop more realistic and well tested scenarios for both water availability (scarcity) and socioeconomic and demographic development in the region, for which there is still not enough experience available. Increasing the capacity for scientific analysis and policy making with researchers and water managers as well as with high level decision makers is also a key to success for this process. Last but not least, in search of future options for solutions realistic trade-offs have to be developed on the basis of high level planning discussions with regional and international stakeholder participation.

Based on the present institutional framework described in this paper it is suggested that the ICWC take the lead in this process with the support of the Scientific Information Center as the process coordinator. In order to properly support the process a wide range of dedicated research and development work needs to be done and SIC should make use of the existing scientific framework of government institutions for water management and monitoring, research institutions, design institutes and contracting companies as much as possible. There is still a lot of communication between the members of this network and much of its research is based on common principles. The longer the start of the project is postponed the more this will cause losses to this common framework and increase the complexity of the process. Moreover efforts should be made to involve Afghanistan as soon as possible.

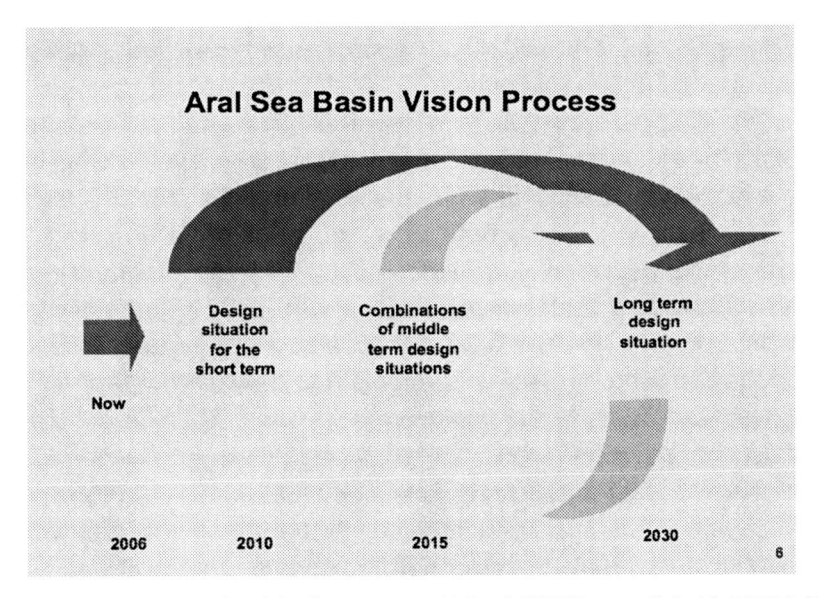

Figure 2. Aral Sea Basin visioning process. (After LTV Western Scheldt 2002.) (7)

The vision process should start with the development of, and agreement on, the vision process agenda including agreement on geographical boundaries, composition of the water resources management issues (building blocks for the vision) and the process of preparation towards the actual process for development of the vision and of the design steps. A first step (as shown above) is to agree on a design situation for the main issues (technical, financial, ecological, legal, and institutional) for the short term in order to be able to let the system function properly under agreed present (temporary) conditions. This will create a time frame (e.g., 4–5 years) for development of the long-term design situation (vision) for the basin.

During this period a wide variety of design issues (technical, financial, ecological, socioeconomic, legal, and institutional) will come under study and will be forwarded for discussion in various stakeholder arenas. Existing models will be improved and new models will be developed and results will be tested both on regional and national levels. This phase should lead to an agreement on the wider framework for a long-term design situation for the Aral Sea Basin among the countries and institutions of the region after communication with the main stakeholder groups affected. It is important at the this stage to have a very thoroughly developed framework for calculation of costs and benefits and come to a set of commonly agreed scenarios and strategies that will lead to the desired sustainable end situation in for example 2030 (next generation).

In this process approach it is important that options for development that fit the overall design framework are tested and evaluated on their performance in view of their contribution to the overall long-term goal. This may take the format of actual pilot projects testing new policies or management approaches directly (e.g., different operational regime for a hydropower plant) or indirectly (farm subsidy for different cropping patterns) related to water use issues. Individual middle term design options or combinations thereof will contribute to the agreed end situation and create room for participants in the process to adjust and gain confidence. The sequence in the process is that first middle term design policies are developed and discussed followed by actual design pilots or combinations of design pilots. The implementtation of the pilot projects starts approximately 4–5 years after the start of the process.

5. Monitoring and Information Sharing

A key activity in support of the process is to produce relevant and credible information and develop the concepts and tools necessary to communicate results to the different target groups. This may involve a large variety of products ranging from specialized training to the development of dedicated and integrated models, role plays, workshops and seminars, and public campaigns. Especially integrated models that can be fed with results of specific research projects may prove to be very important in this respect. One strong example is the Aral Sea Basin Management Model (ASBmm) that was developed with support of UNDP and WB funding. It incorporates hydrological modules and socioeconomic and demographic growth modules that have been developed by experts both inside and outside of the region. The models use data that have been collected by many different institutions in many different projects implemented in the region throughout the years. The model was developed with the objective to enhance awareness among a general public (students, scholars, civilians, water managers, politicians). The model can analyze consequences of different scenarios (population growth, economic development, climate change) in combination with different trans-boundary water management strategies. It is available both in English and Russian.

The model was developed for different user levels in variation of a (widely distributed) base model for a general public to an expert model version used by policy makers, water managers and as a training tool to a third very detailed model that is capable to deal with water management issues on a planning zone level. This last model is used for planning and research work for the SIC-ICWC and river basin management organizations within the region, for which it using the extended databases produced by recent projects sponsored by various financing agencies including EU TACIS, World Bank and other donor organizations. Similar models have been developed by other institutions in different countries of the region and many make use of existing or recently developed spatial and non spatial databases with information on hydrology, hydraulic infrastructure, socioeconomics, land use, etc. It will be important to, from the beginning of the process to have these data properly inventoried and made available for research in the project. Continued development of the ASBmm and similar models by top scientists from the partner countries in the Aral Sea Basin and international experts will be a priority activity of the project. Another important issue is to reach agreement on the methodologies and indicators for basin wide monitoring of water resources (quality and quantity) and water use and water productivity data. An agreement on a common framework for water resources monitoring and data sharing will be part of the long-term design framework for the Aral Sea Basin.

6. Capacity Development

Since the collapse of the Soviet Union much of the capacity for water management and research in the region has suffered from economic decline and lack of funding which has caused many young scientists in the sector to seek alternative occupations. Many experienced scientists of the older generation are close their retirement age and the pace of their replacement is slow and incomplete. This has led to situations where there is now staff with insufficient education and training or inadequate educational background entering the offices of the water management institutions. ICWC members have recognized this process and have started to work on the development of capacity for water management through new education and training programs. For example starting in the year 2000 and with the help of Canadian CIDA, SIC ICWC has established a training centre network with a hub in Tashkent and branches in the Central Asia partner countries. The Training Centre has developed and organized water and environmental management training courses for over 1000 participants and has promoted many new programs on the sharing and managing of transboundary water resources in the context of regional cooperation. Besides being a provider of specialized training the Training Network of SIC ICWC also functions as a kind of "round table" for discussions on the issue of regional water management and has transformed itself into a think-tank for regional collaboration.

From the problem analysis and consequently identified need for increased capacity for the water sector described in this paper it is clear that SIC-ICWC is facing increasing pressures due to its functions and responsibilities. This requires the institute to design and implement dedicated water resources training and capacity development activities that will enhance capacity for development of sustainable management practices that are now urgently needed in the region. The recent cessation of financial support for the Training Network from existing international sources has created a challenge for continued capacity building for the water sector of Central Asia. The water planners of Central Asia need to act urgently in order to ensure long-term water security for their population and avoid conflicts about water in the future. An additional condition to the success of the proposed visioning process is that SIC-ICWC, possibly in coordination and cooperation with other partners in the region, will seek new opportunities for organization of capacity building for integrated water resources management (IWRM) expertise for Central Asia over a number of years. One specific option that has been developed in this respect is a joint proposal by UNESCO-IHE and SIC-ICWC for Training and Capacity Building for Water Resources Planning and Management for Central Asia and Afghanistan. This proposal foresees a cooperative program for education and training for water management practitioners in Central Asia with emphasis on the Aral Sea Basin countries, including Afghanistan. Two separate education and training options are proposed to be provided over a limited number of years:

- A Master's Program, focusing on IWRM, where participants from Central Asia and Afghanistan participate in the regular MSc. programs of UNESCO-IHE

- A Training Certification Program in order to provide flexible options for prospective Central Asian students that require tailor made training programs. These training initiatives may be organized abroad or in the region as specific needs and possibilities are identified

The proposed training modules is to be further refined based on feedback from Central Asian and Afghan field experts once the project is approved and has obtained financial support. Including Universities and research centers of the participating countries of the region will add to the effectiveness of the program and create opportunities for increasing institutional capacity.

References

Bosnjankovic B.; Negotiations in the Context of International Water Related Agreements, UNESCO/IHP/WWAP, PC>CP Series, 8, 2003

Dukhovny V.A. and Sokolov V.; From Potential Conflict to Conflict Potential: Lessons on Cooperation Building to Manage Water Conflicts in the Aral Sea Basin, UNESCO/IHP/WWAP, PC>CP series 11, 2003

Dukhovny V.A. and Deschutter J.L.G.; South Priaralie New Perspectives: final report of the NATO SfP 974357 project, Tashkent 2003

SIC-ICWC, Mountain Unlimited, NES, SIBICO, DHV; Economical Assessment of Joint and Local Measures for the Reduction of Socio-Economical Damage in the coastal Zone Areas of the Aral Sea, INTAS 1059, Tashkent 2004

SIC-ICWC, Resource Analysis; The Aral Sea Basin Management Model (ASBmm) for awareness raising and strategic decision support: final report and integrated river basin model, Tashkent–Delft 2002

SIC-ICWC, CWSIR; REBASOWS final report on the rehabilitation of the ecosystem and bio-productivity of the Aral Sea under conditions of water scarcity, INTAS 0511, Tashkent–Vienna 2006

Resource Analysis; Long Term Vision for the Scheldt Estuary final report, Netherlands–Flanders ProSes Project Bureau, Bergen op Zoom 2002

Bogardi J. and Castelen S., Selected Papers of the International Conference on Co-operation in Integrated Water Resources Management: Challenges and Opportunities, UNESCO-IHP/IHE, Delft 2003

WATER ECOSYSTEMS OF CENTRAL ASIA: IMPORTANT FACTORS AFFECTING THE ENVIRONMENTAL & SOCIAL PROSPERITY OF THE REGION

Water is a critical factor for the archebiosis on the Earth, and it remains the key element for surviving. Preserving inland waters is the basis for supporting all important goods and services supplied by aquatic ecosystems. ("What is a Water?...." The official publication of the Secretariat of the Ramsar Convention, 2005).

BULAT YESSEKIN, MALIK BURLIBAYEV,
NINA MEDVEDEVA, AND SANZHAR STAFIN
*The Regional Environmental Centre for Central Asia
(CAREC)
Almaty, Kazakhstan*

Abstract: The historical approach of valuing water in Central Asia based soley on proction quotas or financial contributions to industry has led to a worsening of regulatory, supply and supporting functions of aquatic ecosystems in the Central Asian and the Caucasus region. The welfare and future prosperity of the countries of this region are linked to the status of aquatic ecosystems. An overview of the issues related to water resources management in the region, the programs adopted in the region for preservation of aquatic and water-related ecosystems, and the role of the international community in the protecttion of regional ecosystems from degradation is provided. Despite the persuasive example of the Aral Sea disaster, the use of water resources is basically planned to meet the water requirements of agriculture and hydropower generation without consideration of other needs of the economy and nature. As a result, drinking water quality and health of local population are deteriorating; land productivity and crop yields are decreasing; and poverty, unemployment, migration, and risks of conflicts are generally on the rise.

Keywords: aquatic ecosystems, Central Asia, biodiversity, sustainability, water resources management

1. Introduction

Sustainable development depends on the continued ability of ecosystems to support environmental parameters vital to human health (favorable climate, optimal composition of atmospheric air, safe water, and foodstuffs) and for economic and social development. Aquatic ecosystems (AE) play an important role in maintaining the welfare of nations and in preserving biological diversity in sub regions of Central Asia and the Caucasus.

However, the resource consuming activities that are associated with economic activity along with insufficient consideration of ecosystem water needs to allow them to sustain often leads to the destruction of regulatory, supply, and supporting functions. Ongoing degradation of the Aral and Caspian seas, reducing biodiversity and biological resources, and adverse changes in transboundary river flows are widely known. "These processes result in deteriorated drinking water quality and health of the population."[1] The vulnerability of aquatic ecosystems in Central Asia (CA) is a major limiting factor for sustainable social and economic development.

The countries of CA face the acute need of developing and implementing the integrated actions aimed at settling the increasing problems of destruction of AE. The United Nations (UN) Declaration on Environment and Development (Rio Declaration), UN Sustainable Development Program "Agenda 21", UN Millennium Declaration and other international documents consider the protection and preservation of ecosystems as the integral part of the developing process. The UN Millennium Declaration adopted by 147 Heads of State and Government and 189 nations at the UN Millennium Summit, sets forth the principles of sustainable development and declares their firm intention to adopt in all our environmental actions a new ethic of conservation and stewardship. The Millennium Ecosystem Assessment Report confirms that ongoing degradation of ecosystems is the major obstacle on the way of achieving the Millennium Development Goals. At the All-European Ministerial Conference held in Kiev (2003), the preservation

---

[1] http://www.unece.org/env/documents/2003/ece/cep/ece.cep.106.rev.1.e.pdf

of aquatic ecosystems was declared as the priority sub regional goal (Goal 1) in Central Asia.[2] In accordance with these goals, the CAREC under support from the Global Water Partnership has studied the status of aquatic ecosystems in Central Asia based on available information. This report will promote public awareness with respect to the degradation of aquatic ecosystems in the sub region, as well as formulating the topical tasks in the field of preserving aquatic ecosystems in Central Asia, and development of efficient strategies and mechanisms for regulating their vital functions.

2. The Central Asian Region

The Central Asian region is located in the center of the Eurasian continent and occupies the area of 3,882,000 km^2 with a population of 53,000,000 people. It includes the countries as Kazakhstan, Kyrgyzstan, Tajikistan, Turkmenistan, and Uzbekistan. As a whole, the physical and climatic zone is well defined in Central Asia. An arid climate that predetermines the vulnerability of ecosystems is the distinguishing feature of this region. Countries of the region are situated in the closed basins of the Caspian Sea, Aral Sea, Lake Balkhash, and Lake Issyk-Kol and have no connections with the world's oceans. In combination with the arid climate, this aspect places constraints on future economic activity and trade.

The status of the aquatic ecosystems in the greater part of the sub region is far from optimal. Because of various economic activities, most aquatic ecosystems have been transformed. Changes can be observed in all natural-climatic zones, and are gradually increase toward the lower reaches of rivers and areas of high river flow consumption.

In Central Asia, water resources in the mountain regions are used for irrigation and water supply to lowland regions. Particularly large irrigated areas are concentrated along the middle and downstream reaches of the rivers Amu Darya, Syr Darya, Zeravshan, Talas, Naryn, Ili, and Chu, as well as on the foothill plains. Natural conditions are favorable to cattle breeding on grazing lands, and, at the same time, the dense river network and plenty of man-made water bodies create conditions for developing commercial fishery. In the region, since the second half of the 1990s the growth in the agricultural industrial

---
[2] http://www.unece.org/env/proceedings/html/Item7b.e.html

complex has been experienced (in Kyrgyzstan–by 9%; in Tajikistan–by 4%; in Turkmenistan–by 26%; and in Kazakhstan–by 29%); as well as an increase in agricultural crop production (in Turkmenistan the area under grain crops is increasing; in Uzbekistan–areas under orchards, melons and gourds, vegetables).

The sustainability of river ecosystems in the sub region mainly depends on the natural equilibrium in upper watersheds (mountain systems of Pamir-Alai, Tian Shan). Along with the growth of urban population and the developing tourism and recreation industries, the use of mountain areas for recreational purposes has been increasing. Therefore, the need to control recreational loads on mountain ecosystems is important. The processes of deforestation, soil erosion, degradation of pastures, and pollution by wastes in mountain areas of both regions are strengthening. Destruction of mountainous ecosystems results in a loss of their regulatory functions and the increase in risks of natural disasters (e.g., floods).

Rise in air temperatures and contamination of precipitation causes increases in the melting of glaciers. This increased melting has already resulted in disturbances of the hydrological regimes, depletion of water resources, and degradation of aquatic ecosystems, especially in the runoff dissipation zone.

A considerable portion of the aquatic and water-related ecosystems of the region is located in foothill areas and adjacent valleys. The development of the mining, processing and chemical industries, as well as urbanization of these territories entailing the disposal of insufficiently treated industrial and domestic wastewater into water bodies are a threat to the functioning of ecosystems. Pollution of the territories and watercourses by wastes is a major problem in both regions and this results in enormous economic and ecological losses. According to monitoring data, 6–8% of water bodies in the area adjacent to the Aral Sea are very polluted with poor ecological conditions: 25% – polluted; 44% – moderately polluted; and 23% – clean and slightly polluted.

Irrigation and developing of the plain areas in Central Asia and the Caucasus have caused considerable transformation of aquatic and water-related ecosystems. Not only is there a loss of some components of biodiversity (species of flora and fauna) but also the extinction of whole ecosystems. As a result of the transformation of aquatic and water-related ecosystems and direct exploitation, more than 50 species of fish, about 40 species of birds, 20 species of mammals, 4 amphibians many of which were included in the IUCN List are now referred to as threatened species.

Construction of reservoirs has contributed to the destruction of ecosystems in the lower reaches of rivers and has adversely affected the long-term prospects of economic development of countries and regions. River flow regulation has resulted in (1) a decrease in volumes of biologically active sediments (2) changes in the hydrological regime of downstream reaches of watercourses (3) increased sedimentation of riverbeds and (4) disturbance of the migration of migratory fish. Reservoir construction has also resulted in changes to the bottom sediments of rivers. More coarse, sandy sediments bury silty bottom sediments downstream from reservoirs. Aquatic pelophilic organisms fail and perish because they cannot properly feed and build their shelters on the altered bottom sediments.

The situation in the deltas of the Amu Darya, Kura, Syr Darya, Razdan, and Ural rivers is at an emergency state. There is a reduction in the volume of intra-delta water bodies, an increase in their water salinity levels, and a decline in biodiversity and bio-productivity of the deltas as a whole, as well as losses and a reduction in the number and diversity of natural habitats. Fishery conditions in the Syr Darya delta have changed dramatically with 97% of spawning grounds becoming inaccessible for barbel sturgeon (*Acipenser nudiventris*), 95% for barbel (*Barbus barbus*), and 60 percent for asp (*Aspius aspius*) and white-eye (*Abramis sapa*). The area of fishery lakes in the Syr Darya River delta has decreased 12-fold when compared with the 1960s. In the last ten years, fish catches from natural water bodies have declined by more than by 60% in the Central Asian region as a whole.

Intense sedimentation, transformation of coastlines, and changes in the hydrological regime has resulted in a decrease of the self-purification capacity of reservoir aquatic ecosystems. Intense eutrophication and secondary pollution of the shallow zones are the typical problem for many in-channel reservoirs in the zone of the plains due to negligible depths and large areas. The removal of silt from riverbeds resulting from the regulation of river flows has proven to be expensive.

Natural lakes of the sub region contain the considerable reserves of fresh water, and their ecosystems have a regulatory influence upon climate and the functioning of other ecosystems in the basin. The largest of these lakes are: the Aral Sea, Lake Issyk-Kul, Lake Balkhash, and Lake Sarez. As a result of reducing river flow, many rivers, especially small rivers, have brackish water. Decrease in the habitat area of colonial-nesting species of water-related birds (herons, cormorants, and pelicans) is in progress, and the conditions of the winter habitat of an enormous number of waterfowl during their migration from the North Europe, Siberia, and Kazakhstan to wintering grounds on the Caspian Sea, in India, Pakistan, and Africa are deteriorating.

Due to the irrigation schemes, an area of desert sinks that accumulate enormous reserves of drainage water has increased in Central Asia. According to remote sensing data, more than 300,000 ha of desert grasslands were flooded with drainage water in Central Asia in the 1970s and 1980s, and 800,000 ha in 2005 in Uzbekistan alone. The percentage of largest desert sinks–Lake Sarikamish and the Arnasay Lake system–makes up 70% of the area of such water bodies. New-formed aquatic ecosystems play an increasing role in the water balance and landscapes of desert areas. The ecological effects in these areas are different. Desert sinks became the peculiar ecological oasis– the zones for supporting biological diversity, and at the same time, they are involved into the socioeconomic sphere and are used by the population for recreation, fishery, hunting, harvesting reed, etc. However, such processes as their intense overgrowing with wetland vegetation, shallowing, gradual salinization, and hydrogen–sulphidous contamination of near-bottom (benthic) layers occur here. In water-related ecosystems, valuable forage plants are replaced by less-valuable species (over an area of about 530,000 ha).

There is the need to manage the water quality of desert sinks with the purpose of increasing their biological productivity. Unfortunately, up to now, these aquatic ecosystems do not have the economic and ecological status, and the legislative documents establishing the basis for their management were not adopted. Fishery and recreational value of these water bodies, resources for hunting, animal breeding, and other kinds of economic use of their biological resources should be assessed. The problem of return water and numerous water bodies that were created owing to return water disposal also needs to be considered both at the national and regional level. It is necessary to specify and legally fix the social and ecological status of desert sinks.

Ecosystems of man-made watercourses (irrigation and drainage canals) play an important role in the functioning of the river basins. In the countries of the sub region, these areas make up hundreds of thousands of square kilometers, and their length consists of many thousands of kilometers. For instance, only in Uzbekistan, the main irrigation and drainage canals cover an area of 156,000 km^2, and their total length is about 1,100 km. Water delivery and use are often complicated deteriorating water quality and sanitary conditions. Therefore, development of measures for addressing these problems has a practical value and is one of the most important issues of hydrobiology.

The lack of agreements with respect to water sharing among countries and economic sectors (basically between irrigated and hydropower generation sectors), as well as the nonobservance of signed agreements has resulted in decreasing water availability for economic sectors in downstream districts, decreased irrigated areas and output of agriculture, and in deterioration of ecological conditions in lower reaches of the rivers. Excessive water consumption for irrigation in the middle parts of river basins causes a shortage of water supply to aquatic ecosystems in the lower reaches. The general problem for both regions is the inability of the existing management system and water infrastructure to provide the ecological flow for supporting of spawning grounds and ecosystems in river deltas. In many countries, the rates of ecological flow, sanitary and sanitary-ecological water releases are not estimated and depend on annual water availability. In practice, water supply for maintaining ecosystems is often formed according to the residual principle especially in the drought periods. In the dry 2000–2001 years, there was a problem of meeting drinking water needs in lower reaches of the Amu Darya River. During the period between the consecutive surveys of lakes (since 1936 until 1985), an area of lakes hydraulically linked with rivers has decreased from 380,000 to 30,000 ha in the flat part of the Aral Sea basin as a whole, and in the Amu Darya delta this area has declined in area by 7 times. Aquatic ecosystems support vital functions of other ecosystems in the delta and also function as an integral natural complex, especially under arid conditions. The regime of water delivery into river deltas in place in the last several years is ineffective for supporting the biological productivity of their aquatic ecosystems. Prevailing of winter water releases cause the problems of winter floods in the deltas and abatement of their fauna. The destruction of aquatic ecosystems in the deltas has affected not only the biological diversity but also the local population whose incomes mainly depend on usage of biological resources. Deteriorating the environmental condition in the Amu Darya River delta has affected the interests of 1.5 million people living in the region. Risks of undermining the food independence and safety are increasing in the countries of the sub region.

Alien species such as ctenophores (*Mnemiopsis ieidyi*) that were brought with ballast water of oil tankers and have invaded the Caspian Sea adversely affecting the marine ecosystems. Sustainability of aquatic ecosystems and their self-purification ability are defined by the condition of phyto- and zoocenosis, whose composition and structure directly depend on hydrological and hydro-chemical regimes. Insufficient studies of aquatic biocenosis and lack of routine monitoring of surface water pollution do not allow for an in-depth analysis of the reaction of the biological component of aquatic ecosystems on the increasing pollution in the sub region. This matter requires additional research and the establishment of a biological monitoring system.

Water pollution is a problem for all the countries of the sub region. Drainage water disposal, discharging of agricultural and industrial wastewater, and the discharge of pollutants within the water protection zones along rivers contribute to increasing chemical and bacterial pollution of water resources. River water quality in the zone of intense water consumption varies mainly from Class III to Class IV (polluted water), and in the zone of large-scale industrial and urban complexes water quality can drop to Classes V and VI (dirty and very dirty waters) when the water is ecologically hazardous with clear degradation of aquatic biota and cannot be used for any purposes. Due to the pollution of water from these sources, increases in morbidity of the population and deterioration of the quality of agricultural output are observed. The major watercourses of Uzbekistan became unsuitable for drinking water supply due to a lack of freshwater sanitary and ecological releases. The largest problem in the Caspian region that results in enormous economic and ecological losses is pollution of coastal-marine ecosystems by oil products. Lack of both urban (rural) and local treatment facilities and low efficiency of existing plants is evident. The need to regulate industrial production and treatment methods for wastewater should be met, but economically feasible solutions are very often ecologically not found to be acceptable.

The environmental problems such as water resources shortage, pollution and degradation of aquatic ecosystems directly affect the economic development and other social issues including poverty, forced migration, deterioration of life quality and health of the population in the CA region. Medical surveys conducted in the 1990s in Central Asia have revealed an increase in incidences of diseases of the following organs: endocrine and urino-genital system, digestive apparatus, blood and hemopoietic organ, circulatory system, as well as oncological diseases due to deteriorating quality of natural waters. Evaluating dynamics of health of the population in areas adjacent to the Aral Sea such nosologic forms as morbus hypertonicus, stomach ulcer, and duodenal ulcer confirms a leading role of general water salinity and its salt content composition in their etio-pathogenesis can be observed. For example, under increasing concentrations of chlorides of up to 50 mg/l, the morbidity rate related to cholelithiasis and cholecystitis rises threefold, and for stomach ulcers–almost fourfold.

All countries recognize that the top-priority task for settling key problems of the safety of water resources and the environment is water supply/quality to the population, economic production, and ecosystem quality.

3. Developing Efficient Strategies and Mechanisms for Ecosystems Management

Important causes leading to the previously mentioned problems of aquatic ecosystems degradation in the sub region is inefficient management and lack of public concern. In the different periods of the Soviet epoch, the integrated water resources use and protection plans, integrated nature protection plans, municipal economy development plans, and water infrastructure and land reclamation construction plans were prepared according to a standard methodology. They did not sufficiently consider sustainable and equitable needs of water users and nature, as well as the balance among different natural and economic zones.

At the same time, the neglect or insufficient consideration of ecosystem-defined limitations has resulted in a crisis in nearly all of the river basins of the sub region. Society often perceives the preservation of ecosystems as a low priority which is only the concern of nature protection agencies. There is not still the legal status of the term "preservation of ecosystems" in the normative documents of countries of Central Asia, and agencies responsible for the preservation and maintaining of ecosystems within the framework of the state governance were not specified as well[3].

"Governance of the water sector should be modernized in order to equally represent the interests of irrigation, hydropower generation and other stakeholders while observing the priorities of drinking water supply, water-saving etc. and ensure the principle of equality of rights and responsibilities of all water users". (The invitation to the partnership for implementing of the Central-Asian initiative on sustainable development, 2003). Such modernization is possible utilizing integrated water resources management (the Global Water Partnership) based on the following principles of such an ecosystem approach:

- Water resources management within hydrographic boundaries
- Managing all water resources: surface water, groundwater, and return water while recognizing their interactions
- Integration of the interests of all economic sectors, water users, and water consumers
- Integration of the different levels of management hierarchy
- Public participation in the decision-making process
- A recognition of nature as a water user

The grave obstacles for implementing the basin management approach into practice are the following:

- The priority of short-term tasks in the governmental and territorial planning and the activities of private businesses
- Lack of the programs for integrated development of regions comprising the economic component
- Lack of enforcement of existing environmental laws
- Limited financial and institutional capabilities of ministries
- Weakness of civil society and NGOs for defending an opinion of local communities and the needs of ecosystems

---

[3] The Ministries of Nature Protection are responsible for many aspects related to nature protection (from the wastes disposal control to collection of penalties for non-observance of ecological laws), but functions directly related to the preservation of ecosystems were not exactly specified. Therefore, the agencies responsible for nature protection do not still solve the matters related to specifying and control of water releases for the needs of ecosystems.

- Insufficient efficiency of available mechanisms for settling disputes between different water users comprising the transboundary level
- Lack of incentives for water saving
- Lack of an ecological-economic assessment of the current status of aquatic ecosystems
- Neglecting the protection of ecosystems under developing recreational areas
- Lack of appropriate incentives for the application of the resources-and-energy saving technologies
- Lack of a national monitoring systems to better understand and control ecosystems and resources

4. Measures for the Preservation of Aquatic Ecosystems

In the countries of the sub region, the following projects and programs aimed at improving the environment were implemented or are in process of being implemented: the Concept of Water Sector Development and Water Policy of the Republic of Kazakhstan up to 2010 (Kazakhstan), Kyrgyzstan Integrated Water Resources Use and Protection Plan up to 2005, Irrigation and Water Supply Infrastructure Rehabilitation Programs (Republic of Tajikistan), the National Environment Action Plan of the President of Turkmenistan Saparmurat Turkmenbashi.

In the programs adopted in the countries of the sub region, the preservation of aquatic and water-related ecosystems is the top-priority task for national sustainable development. The following actions are planned to be implemented:

- Improve the legislation and strengthen the control for observance of laws
- Develop national action plans in the field of integrated water resources management (including the ecosystems supporting them)
- Introduce water-saving technologies and achieve the minimum level of available water resources losses
- Specify the scientifically-grounded rates of ecological river flows and measures for improving the condition of aquatic ecosystems

- Control river water pollution and water-related disasters
- Develop activities aimed at accounting, reproduction, and the growth of biological resources
- Rehabilitate vegetation cover in the zones of runoff formation and consumption
- Control mudflow and flood control by means of construction of protective levees and cleaning of silted riverbeds
- Maintain activities related to the protected areas
- Establish water protection zones along rivers and other water bodies
- Improve the management of the coastal zones
- Enhance the existing water quality monitoring system and the control for discharges of pollutants into aquatic ecosystems
- Develop methods for evaluating damage of aquatic ecosystems and the rates of compensations for polluting and depleting water resources of transboundary rivers in line with the international standards
- Develop quantitative criteria for evaluating the interrelations of living standard of the population and the status of ecosystems
- Define social, economic, and ecological values of water resources and
- Promote public awareness and encourage public participation in these processes

Assessment of the health of the aquatic environment was recently introduced into the practice of nature protection. A forecast of aquatic ecosystems status and development of the scenarios of their modifications are extremely important for long-term planning for a water body. Taking into consideration the experience of activity within the framework of the Ramsar Convention (1971), the tasks related to the protection of biodiversity of aquatic ecosystems in the sub region should integrate the following aspects in order to:

- Facilitate sustainable water resources use in all economic sectors including optimization and introduction of new systems of irrigation and water supply (the Millennium Development Goal 7,

Target 10: Indicator 30 Proportion of population with sustainable access to an improved water source, urban and rural)

- Develop a system of measures for the management and protection of aquatic ecosystems in order to establish protected areas presenting an optimal network for supporting biodiversity based on auditing of all aquatic ecosystems and specifying the key sites, as well as on routine monitoring and management of such a network (the Millennium Development Goal 7, Target 9: Indicator 26 Ratio of area protected to maintain biological diversity to surface area)

- Strengthen public awareness with respect to the values of aquatic ecosystems and their biodiversity by means of economic incentives for enhancing commitments of the local population and authorities regarding the preservation of biodiversity.

Protection and rehabilitation of forests, overgrazing control, measures for the prevention of mudflows, landslides, catastrophic floods due to breaches of mountain lakes, control of chemical and biological pollution, and quantitative depletion of water resources are the basic activities related to the preservation of terrestrial and aquatic ecosystems in the runoff formation zone. It is necessary to develop and introduce legislative and institutional measures for controlling tourism in order to regulate the recreational load on vulnerable mountainous ecosystems for preventing pollution of water sources and reducing the biological diversity by campers and others.

The basic shortcomings of the existing integrated water resources use and protection plans, integrated nature protection plans, municipal economy development plans, and water infrastructure and land reclamation construction plans is that they do not take into consideration sustainable, equitable, and wise water delivery to ecosystems. Some actions in this direction have already been undertaken–the Aral Sea is recognized as a water user equal in rights, however, it is necessary to implement this resolution in practice. The integrated aquatic ecosystem use and protection plans with considering the ecosystems of the Aral, and Caspian Seas should be developed for the protection of aquatic ecosystems of transboundary rivers. For many national and international experts, the process of preparing the agreements concerning specific problems of interstate water relations can be accelerated by adopting the general strategy of efficient use and protection of terrestrial and aquatic ecosystems, as well as establishing and strengthening the status of the sub regional coordination commissions.

Implementing the pilot projects for rehabilitation and sustainable management of aquatic ecosystems in different natural and climatic zones with wide dissemination of their outcomes and experience would also be useful.

4.1. LEGISLATION

Some actions for improving the water legislation were undertaken for the first time by the countries of the sub region. The new Water Codes (Kazakhstan, and Kyrgyzstan), that initiates the development of the national water program and basin management plans were adopted recently. At the same time, there is a discrepancy between the normative documents and legislative norms that causes conflicts among water management bodies and water users and adversely affects the procedure of water resources management. There is still not the legal status of the term "preservation of ecosystems" in the normative documents of the countries of Central Asia.

Taking into account the requirements of the international conventions, further improvement and coordination of the normative and legal base of water relations and harmonization of the legislative base are necessary. The process of ratifying and executing UN Conventions and Protocols (The Convention on the Protection and Use of Transboundary Watercourses and International Lakes [*the UNECE Water Convention*], the UNECE Convention on EIA in a Transboundary Context etc.) and signing the basin agreements and memoranda needs to proceed.

All basin agreements related to water resources integrate and recognize the behavior of riparian countries promoting the presservation and protection of ecosystems, integrated water resources management, and reduction of diseases caused by poor water quality. The commitments of riparian countries with respect to unilaterally planned water use, procedures of transboundary environment impact assessment, and distribution of responsibilities in case of floods, droughts or emergencies should be addressed. In addition, agreements should be established regarding the order of consultations and effective mechanisms for notification, control, and mitigation of transboundary impacts, including detection of pollution sources, measures for reducing water pollution, and monitoring water quality. At the same time, the agreements have to include the mechanisms for reducing the risks to public health caused by poor water quality, and providing effective mechanisms for public awareness and participation, as well as addressing obligations with respect to damage and dispute settlement mechanisms.

4.2. INSTITUTIONAL BASE

The governance system for the water sector in the countries of the sub region comprises the large number of ministries, departments, and organizations with an extremely complicated framework for coordination that obstructs establishing and developing the integrated management of aquatic ecosystems. The responsibility for the status and preservation of ecosystems is not specified in the system of state governance in the countries of the regions[4]. The system of basin governance is at the initial stages of development (Kazakhstan, Kyrgyzstan, and Uzbekistan).

Water supply, protection of water resources, and preservation of aquatic ecosystems are not only national problems but also largely depend on the coordinated relations of riparian countries in the sub region. Experience gained in Central Asia shows that despite the efforts of the Interstate Coordination Water Commission (ICWC) at the regional level who adopted political resolutions and agreements in the field of efficient natural resources use and nature protection, such problems as the insufficient legal and institutional base for regional cooperation, lack of political and financial obligations, and insufficient participation of riparian countries in tackling regional challenges. A number of programs, declarations, and agreements adopted by the Heads of Central Asian States with respect to developing the strategy of water sharing acceptable for all riparian countries and the economic mechanisms for transboundary water resources management were not still actually realized.

---

[4] The Ministries of Nature Protection are responsible for many aspects related to nature protection (from the wastes disposal control to collection of penalties for non-observance of ecological laws), but functions directly related to the preservation of ecosystems were not exactly specified. Therefore, the agencies responsible for nature protection do not still solve the matters related to specifying and control of water releases for the needs of ecosystems.

The common interests and targets regarding the preservation of aquatic ecosystems and improvement of water resources quality are the base for the further inter-sectoral and inter-state cooperation. Central Asian countries have proposed the partnership initiative for sustainable developing of the region ("Agenda 21") that was included into the resulting documents of the UN Conference on Environment and Development (UNCED). This initiative provides for the integration of real processes and strengthening the mechanisms for the cooperation between economic sectors, countries, and donors for achieving general development goals including the goal "Ensuring sustainable functioning of aquatic ecosystems important for vital activity of human beings (Preventing the degradation of aquatic ecosystems in river basins that support vital activity in Central Asia)."

4.3. ECONOMIC MECHANISMS

The countries in the sub region do not have a well-grounded assessment of the value of aquatic ecosystems. Therefore, there is not also sufficient support for the need for ecological flows which would provide for the optimal functioning of natural and man-made ecosystems. In recent years, the natural complexes in most of countries of the region received water in volumes earlier than specified by the basin water management plans and other project studies because of an economic depression, especially in the agricultural sector, and to relatively large amounts of natural water available at this time.

Under market relations, when the economic mechanisms play a basic role, the efficiency of water resources use and indicators of water saving will be mostly depend on the economical assessment of the cost of ecosystems' services and the payment for water resources and services, and at the same time, the economic mechanisms defining water saving and efficient water use will prevail over other ones. The economic and ecological status of aquatic ecosystems including ecosystems of man-made water bodies and watercourses will have to be established.

4.4. PUBLIC PARTICIPATION

In Central Asia, the extent of public participation in decision-making related to water resources management has been negligible due to a lack of awareness of existing problems, limited access to information, and insufficient opportunities for stakeholders to participate in the decision-making process. In line with the 12th principle of the ecosystem approach[5] and the provisions of the Arhus Convention, all interested groups of society and representatives of all branches of science should be involved in water resources management. Measures for building-up of public opinion and support to implementation of the IWRM principles and the preservation of aquatic ecosystems should be addressed.

Participation of water users, local communities, and NGOs in water resources management and the preservation of ecosystems are possible through the Basin Water Councils that should be established within the framework of the basin agreements including the international conventions. One of the first Basin Water Councils was established in the Balkhash-Alakol River Basin (Kazakhstan). However, considerable work still needs to be implemented to enable it to carry out the abovementioned tasks.

4.5. INFORMATION AND MONITORING

The existing monitoring system for ecosystems in the countries of the sub region is not sufficiently effective and does not meet the needs of comprehensive assessment of the status and dynamics of aquatic ecosystems in order to make objective-management decisions.

The process of management decision-making will promote a unified accessible expert-information system comprising not only data on the status of aquatic ecosystems and driving factors but also data on the entire complex of economic, social, technical, cultural, and other processes that affect the basin. Such a system will provide decision-makers with an effective and science-based instrument for developing the options of administrative and legal decisions aimed at ecologically sustainable development of river basins and improving the living standard of the population in the sub region, as well as the prevention of water conflicts.

---

[5] http://www.biodiv.org/decisions/default.asp

5. General Conclusions and Problems

The integrated analysis of the problems of preserving the ecosystems in the Caucasus and Central Asia reveals many causes of their origins and cause-and-effect relations that have general and specific characteristics/features at the national and regional level. The intensity and nature of adverse processes arising in aquatic ecosystems also depend on their location within the river basin (upper watershed, the middle reach of a river or delta).

Causes:

- Lack of the awareness with respect to the significance and role of aquatic ecosystems for preserving the ecological sustainability and supporting welfare of the population
- Lack of integrated water resources management
- Neglecting the protection of aquatic ecosystems (the protection of aquatic ecosystem is not incorporated into the water use system and the appropriate laws and tasks of governmental bodies)
- Enormous water losses exceeding the standard rates in irrigated farming
- Disregard for the water requirements of ecosystems. The rates of ecological flows and sanitary water releases are not specified and mainly depend on annual water availability, not meeting the needs of ecosystems in lower reaches of rivers
- Insufficient use of economic mechanisms, lack of economic evaluating the cost of ecosystems' services and a monetary equivalent of damage due to degrading quality of ecosystems
- Lack of incentives for water-saving practices
- Degradation of forests and pastures, reducing of glacial areas, a threat of breaching mountain lakes, as well as intense erosion over catchment areas
- Untreated discharges of industrial, agricultural, and communal wastewater
- Poaching and introduction of alien organisms
- An underdeveloped analytical and research basis and insufficient knowledge with respect to the processes of forming, man-caused

transformation, and spatial-temporal variability of aquatic and water-related ecosystems in the region
- Ineffective system for monitoring of water quality and the status of ecosystems
- Regional and national information systems are underdeveloped and do not provide access to information for all stakeholders

5.1. CAUSES OF A REGIONAL NATURE

- Integrated water resources use and protection plans for transboundary river basins with considering the needs of ecosystems (management plans) have not yet been developed
- Insufficient capabilities of countries for harmonization of national water and environmental laws at the regional level, in particular, as applied to transboundary aquatic ecosystems
- Competing interests of different water users, including consumers (hydropower generation, industry, and irrigated farming) within a particular country and in countries located in the common transboundary river basin
- Lack of administrative, economic, and institutional mechanisms for protecting aquatic transboundary ecosystems against pollution and depletion including the mechanism of administrative responsibility and
- Lack of an accessible and integrated information system covering the entire water sector in the region

6. Recommendations

Analyzing the status and causes of degradation of river basins in sub regions allows us to formulate the following recommendations with respect to measures for preserving (rehabilitation and use) aquatic ecosystems in the countries of the Caucasus and Central Asia:
- More clearly define the individual roles and commitments of the governmental bodies, water management organizations and other stakeholders in the field of protecting aquatic ecosystems in the region

- Specifying the needs of ecosystems and minimum requirements to ecological flows in the lower reaches of rivers
- Improving existing water resources management systems based on the principles of integrated river basin management
- Developing action plans for integrated water resources management (including measures for supporting ecosystems)
- Improving and harmonization of the water legislation in countries of the region, and developing by laws of direct action
- The legislative validating of ecosystems entitlement for water and intersectoral procedures for the decision-making process
- The joining of countries (Armenia, Georgia, Azerbaijan, Kyrgyzstan, Tajikistan, Turkmenistan, and Uzbekistan) to the UNECE Helsinki Convention and other UN Conventions that play an important role in the preservation of aquatic and water-related ecosystems
- The legislative validation of the economic and ecological status of desert sinks (the desert depressions used for disposal of drainage water from the irrigation schemes)
- Developing and implementing the procedures for estimating social, economic, and ecological values of goods and services provided by aquatic ecosystems
- Developing procedures for economic evaluation of water resources and introducing water "pricing"
- Estimating the damages to aquatic ecosystems and the rates of compensation for pollution and depletion of water resources of transboundary rivers
- Introducing international requirements for water quality control and maintaining minimum sanitary–ecological flows, as well as the protection of potable water sources
- Rehabilitating forests and pastures, the measures for preventing breaching of mountain lakes in upper watersheds
- Reducing water losses by means of remodeling irrigation systems, improving irrigation techniques, employing the technologies for drainage water reuse and other arrangements
- Establishing the water protection zones along rivers and other aquatic ecosystems

- Encouraging public participation and support the IWRM principles, promoting public awareness with respect to objectives and tasks of the preservation of aquatic ecosystems, and developing special educational curriculums
- Establishing a system of monitoring the environment with indicators of surface water quality and parameters for evaluating the status of aquatic ecosystems that has to be agreed by all riparian countries and
- Developing and implementing one or two pilot projects in each country concerning the preservation of aquatic ecosystems with follow-up dissemination of positive experience and development of the regional projects

7. Conclusions

A distinctive feature of Central Asia and the Caucasus is the vulnerability of ecosystems. Central Asia is located within the single ecological area of land-locked basins of the Caspian and Aral seas, and in combination with the arid climate of the region this imposes additional and severe ecological limitations on economic activity and trade.

Development of irrigated farming in the Aral Sea basin at levels unprecedented in modern history has exceeded the resiliency limits of the ecosystems and has resulted in their destruction. Intense water diversion from rivers for irrigation has caused reductions in the water levels of the Aral Sea by 19 m and a reduction in its volume by 75%. By the end of 1980s, the sea has practically died, and this event caused a number of the following adverse consequences: drastic deterioration of water quality and health of local population, large-scale desertification, soil salinization and water logging, decline in biodiversity and intensification of adverse impacts on climate.

The "cost-is-no-object" approach that was formed as far back as in the period of arms race is still prevailing in water-related activities of the countries of the sub region. Despite the persuasive example of the Aral Sea disaster, the use of water resources is basically planned to meet the water requirements of agriculture and hydropower generation without consideration of other needs of the economy and nature. As a result, drinking water quality and the health of local population are deteriorating; land productivity and crop yields are decreasing; and poverty, unemployment, migration, and risks of conflicts are increasing.

The countries of Central Asia need support from the international community. Such questions as the estimation and inclusion of the value of ecosystem services in economic development decisions from global economic and financial institutions, such as WB, IMF, WTO, and others. The discontinuance of the degradation of natural ecosystems, a suspension of economic use of the remaining untouched natural territories (with possible introduction of the moratorium to their use) also in many respects depends on political decisions at regional and global levels.

The welfare and future prosperity of the countries of Central Asia and the Caucasus will depend on the natural balance and the status of aquatic ecosystems. In this connection it is necessary for countries, the international community and other partners to correct existing efforts and to aggressively pursue effective measures to protect ecosystem from continued degradation.

NATO/CCMS PILOT STUDY MEETING ON TRANSBOUNDARY WATER MANAGEMENT ISSUES IN THE UNITED STATES & CENTRAL ASIA: PROBLEM DEFINITION, REGULATION AND MANAGEMENT

MIKHAIL KHANKHASAYEV AND STEVEN LEITMAN
Institute for International Cooperative Environmental Research
Florida State University, Tallahassee, Florida, USA

Abstract: This paper summarizes the main results of the NATO CCMS Pilot Study Meeting on Transboundary Water Management Issues in the United States & Central Asia (8–10 March 2005, Tallahassee, Florida). The structure of the meeting was focused on comparing how major transboundary water sharing conflicts in the United States (like the ACF conflict) have been managed toward the goal of equitable conflict resolution among the competing jurisdictions with similar conflicts in the Central Asian region.

Keywords: Transboundary water resources, Central Asia, water resources management, decision making process, risk assessment

1. Introduction

The Pilot Study Meeting on Transboundary Water Management Issues in the United States & Central Asia was conducted in March 2005 in Tallahassee, Florida. This meeting focused on one of the most complicated and sensitive issues which has a significant and widespread influence on sustainable development of the Central Asian region: transboundary water recourses management problems. The representatives comprehensive overview on the transboundary water issues in their respective countries and adjoining areas. The speakers from the NATO countries shared their experience on the management of transboundary water basins. In particular, the case study on the management of the Apalachicola–Chattahoochee–Flint (ACF) river basin (Alabama, Georgia, and Florida) was presented and analyzed in detail. Also, the practice of integrated water management (IWM) in Belgium was presented in the context of the CCMS Pilot Study on IWM administered by the University of Antwerp.

This meeting concluded the series of meetings conducted within the framework of the NATO/CCMS Pilot Study on "Environmental Decision-Making for Sustainable Development in Central Asia". The acronym CCMS stands for Committee on the Challenges of Modern Society which was established by NATO in 1969 in order to work on solutions related to social and environmental problems. This NATO Committee does not engage in research activities; rather its work is carried out on a decentralized basis with other organizations, through pilot studies. Subjects for pilot studies cover a wide spectrum, dealing with many aspects of environmental protection and the quality of life, including defense-related environmental problems. The list of publications available can be obtained from the CCMS Secretariat. Additional information on the NATO/CCMS Program can be found at http://www.nato.int/ccms/.

The Pilot Study on "Environmental Decision-Making for Sustainable Development in Central Asia (CA) was initiated in March of 2001. Since March 2001, the Central Asian NATO/CCMS Pilot Study organized the following five meetings[1–7]:

1. "Planning Meeting", 26 February–1 March 2001, Silivri, Istanbul, Turkey

2. Working Group Meeting, 16–20 March 2002, The Belgian Nuclear Research Centre Headquarters (SCK-CEN), Brussels, Belgium

3. Working Group Meeting on "Landscape Sciences", 22–24 September, 2002, Almaty, Kazakhstan

4. Working Group Meeting "Landscape Science and Public Health Issues for Environmental Decision-Making in Central Asia", 17–18 March, 2003, Brussels, Belgium

5. Working Group Meeting on "Water and Health Issues in Rural Areas of Central Asia", 4–5 November 2003, Almaty, Kazakhstan

6. Working Group Meeting "Transboundary Water Management Issues in the United States & Central Asia: Problem Definition, Regulation and Management", 8–10 March 2005, Tallahassee, Wakulla Springs, Florida

In conjunction with this Pilot Study, a NATO Advanced Research Workshop on Risk Assessment as a Tool for Water Resources Decision-Making in Central Asia was also conducted on 23–25 September 2002 in Almaty, Kazakhstan.[8]

The main objectives of this Pilot Study were:

- To learn and analyze the approaches and processes used in the Central Asian Countries for environmental decision-making

- To provide up-to-date information to the Central Asian participants in selected areas, e.g., water resources management, and use of environmental impact analysis, through the NATO/CCMS experts.

This NATO/CCMS pilot study was administered by the Institute for International Cooperative Environmental Research of the Florida State University (Tallahassee, Florida).

2. Transboundary Water Resources

Environmental problems can have a significant impact upon human health, the economic and political stability of countries and regions around the world. Shared watersheds and river systems and atmospheric systems may place the environmental goals of an individual country in conflict with regional goals. Environmental degradation also may generate conflict and instability. This Central Asian pilot study was focused on the environmental decision-making process, i.e., on the decision-making process that could protect and/or preserve the environment, and this way, help to facilitate sustainable social and economic development of the Central Asian region.

The countries of Central Asia declared their independence in the 1990s, and many of them are still in transition from the former Soviet Union "command" style economic system to more free-market economic systems. However, these countries inherited from the Soviet system the administrative/regulatory approach to environmental decision-making. This decision-making method is not effective in the rapidly changing socioeconomic and political environment. The CA countries, facing economic declines, are generally not able to provide sufficient funding to support the administrative/regulatory decision making systems that are needed in these countries. In addition, the "command" style decision-making process often prevents excludes the public and stakeholders from the decision-making process which creates significant problems and barriers for future effective resource allocation.

The concluding Pilot Study meeting on transboundary water resources issues was conducted at Wakulla Springs (near Tallahassee), Florida. The structure of the meeting was aimed at comparing how major transboundary water sharing conflicts in the United States (like the ACF conflict) have been managed toward the goal of equitable conflict resolution among the competing jurisdictions with similar conflicts in the Central Asian region. The two case studies that were used principally to orient discussions were the Tri-State Water Conflict in the Southeastern United States (Alabama–Georgia–Florida or Apalachicola–Chattahoochee–Flint Rivers Conflict). This situation is also referred to as the ACF Conflict. The primary Central Asian transboundary water conflict example that was used to juxtapose against the ACF experience was the Aral Sea issue involving the countries of principally Kazakhstan and Uzbekistan.

The meeting was conducted at the Wakulla Springs Lodge; a facility that is situated at the mouth of the Wakulla River in Northern Florida. The groundwater that feeds into the spring originates from many primary sources via deep groundwater flow patterns. Wakulla Springs is one of the world's largest freshwater springs generating approximately four billion liters of freshwater per day. The spring is contained within the Florida State Park System and is considered to be a pristine and natural ecological site in Florida. The remoteness and ecological beauty of the spring made for an appropriate and productive venue for the meeting.

The importance of Florida's estuaries, including the Apalachicola Bay can be summarized with a few economic statistics. One of Florida's primary economic sectors is tourism, which is largely dependent on its beaches and other coastal areas (i.e., in 2004, approximately $53 billion in GDP and 700,000 jobs were attributable to coastal tourism in Florida). Recreational and commercial fishing generate over $5 billion per year; recreational boating contributes more that $10 billion per year, and over 75% of both gross retail sales and taxable retail sales occur in coastal areas of Florida. The long-term vitality (i.e., sustainability) of these economic/ecological resources is highly dependent on the quality and quantity (i.e., flow) of waters entering coastal areas through Florida's River Systems. Rivers that flow southward from Georgia and Alabama to the Apalachicola Bay (i.e., river systems involved in the ACF Conflict) can have significant adverse impacts on both the ecological health of the Bay system and the Florida economy, if not managed properly and managed from a water basin wide perspective. The following two sections provide summaries of ACF Conflict in the United States and the Aral Sea Demise in Central Asia.

2.1. ACF CONFLICT SUMMARY

The states of Alabama, Georgia, and Florida attempted to negotiate an interstate Water Allocation Formula for the Apalachicola–Chattahoochee–Flint (ACF) Basin during the period from 1997 to 2003. This summary focuses on the lessons learned from the failure of this interstate negotiation process in the United States.

The Apalachicola River is formed by the confluence of the Flint and Chattahoochee rivers. The Flint and Chattahoochee are very different in nature and usage. The Chattahoochee's source of flow is primarily surface water and there are multiple storage reservoirs that allow the basin's water supply to be managed. The Flint River, on the other hand, has a large groundwater flow component and has almost no reservoir storage capacity. Therefore, in the Chattahoochee basin flows can be managed both through supply and demand management, whereas flows in the Flint can only be managed through demand management. The Chattahoochee River typically provides slightly more water to the Apalachicola's flow than the Flint, but at times the spring-fed Flint can make a greater contribution.

The major water diversions on the Chattahoochee River are for municipal supply while the major diversions on the Flint are for agricultural irrigation. The reservoirs in the Chattahoochee Basin are managed for municipal and industrial water supply, electricity generation, water-borne transportation, flood control, and recreation. Relative to flows entering the Apalachicola River, the storage capacity of the reservoirs is very limited. About 65% of the storage volume is in the uppermost reservoir, Lake Lanier. Lake Lanier, however, should be managed conservatively due to its location in the headwaters of the basin and because the reservoir also supports the water supply for the Metro Atlanta area. There is also political resistance to lowering water levels in the reservoir from homeowners and recreational users. The fact that Lake Lanier should be managed more conservatively than the other reservoirs contributed to a perception by other stakeholders in Georgia that the State had a bias toward protecting the interests of upstream stakeholders.

This conflict was seen more as a divergence between upstream and downstream interests in the basin than between the three states per se. From the start there was a disparity between how the negotiation process was designed (i.e., by state borders) and how the interests of the stakeholders involved in the negotiation were divided. This discrepancy was most pronounced in Georgia where most of the basin lies. Many interest groups in Georgia were more aligned with the interests of downstream states while the negotiating positions of Georgia tended to be dominated by the interests of the metropolitan Atlanta area (a major population center in NC Georgia).

2.2. ARAL SEA DEMISE SUMMARY

The environmental problem of the Aral Sea Basin demise is among the worst in the world. Water diversions, agricultural practices, and industrial waste have resulted in a disappearing sea, salinization, and organic and inorganic pollution. The problems of the Aral Sea, which previously had been an internal issue of the former Soviet Union, became internationalized after the collapse of the USSR in 1991. The five new major riparian republics–Kazakhstan, Kyrgyzstan, Tajikistan, Turkmenistan, and Uzbekistan–have been struggling since that time to help stabilize, and eventually to rehabilitate, the watershed.

The Aral Sea was, until recently, the fourth largest inland body of water in the world. Its basin covers 1.8 million km^2, primarily in what used to be the Soviet Union, and what is now the independent Republics of Kazakhstan, Kyrgyzstan, Tajikistan, Turkmenistan, and Uzbekistan. Small portions of the basin headwaters are also located in Afghanistan, Iran, and China. The major sources of the Sea, the Amu Darya and the Syr Darya rivers, are fed from glacial meltwater from the high mountain ranges of the Pamir and Tien Shan in Tajikistan and Kyrgyzstan.

Irrigation in the fertile lands between the Amu Darya and the Syr Darya rivers dates back thousands of years, although the Sea itself remained in relative ecological equilibrium until the early 1960s. At that time, the central planning authority of the Soviet Union devised the "Aral Sea Plan" to transform the region into "the cotton belt" of the USSR. Vast irrigation projects were undertaken in subsequent years, with irrigated areas expanding by over one-third from 1965 to 1988.

Such intensive cotton monoculture has resulted in extreme environmental degradation. Pesticide use and salinization, along with the region's industrial pollution, have decreased water quality, resulting in high rates of disease and infant mortality. Water diversions, sometimes totaling more than the natural flow of the rivers, have reduced the Amu Darya and the Syr Darya to relative trickles – the Sea itself has lost 75% of its volume, half its surface area, and salinity has tripled, all since 1960. The exposed sea beds are thick with salts and agricultural chemical residue, which are carried aloft by the winds as far as the Atlantic and Pacific oceans.

While there are other transboundary water problems in Central Asia, the Aral Sea demise is viewed as the most recognizable problem of this type in the region, if not throughout the world.

3. Conclusions and Recommendations

During the meeting, the representatives of the CA region presented a comprehensive overview on transboundary water issues in the region. From the other hand, the speakers from the NATO countries shared their experience of management of transboundary water basins. In particular, the case study on the management of the Apalachicola–Chattahoochee–Flint (ACF) Basin (Alabama, Georgia, and Florida, USA) was analyzed in detail. Also, the practice of integrated water management in Belgium was presented as a general.

There is strong interdependence among the CA countries related to using shared water resources. The principal uses of non-potable water in the region are irrigated agriculture and hydropower generation. While the "upstream" countries, Kyrgyzstan and Tajikistan, generally use the rivers for hydroelectric power generation, the downstream countries, Kazakhstan, Turkmenistan, and Uzbekistan, use water mainly for agricultural purposes. These uses are distinctly different in their temporal and spatial requirements, which in turn create disputes and tensions between the upstream and downstream countries. Also, there are many current issues related not only to water quantity, but to diminished water quality which is transferred by shared rivers Central Asia.

Despite some stabilization and economic growth, there are inter-country disagreements concerning methodological approaches, the monitoring of the environment, quality standards of the environment, the use of appropriate norms and regulations guiding usage of natural resources and the management of shared river systems. The data presently available on water quality and quantity are generally not sufficient for effective decision-making. Resulting ecosystem deterioration has also led to a noticeable reduction of biodiversity in the river basins, as well as indirect hazards posed by adverse impacts on water bodies such as the Aral Sea (e.g., airborne particulates with associated toxic components as a result of desertification). There is an urgent need to prioritize and better manage the activities in these river basins leading to preservation of ecological systems in the Central Asia region.

The case study discussion conducted during this Pilot Study meeting was focused on the legal, technical/modeling, public participation, and regulatory aspects of transboundary water issues. The comparison and contrast of the ACF situation with similar situations in Central Asia, including the Aral Sea demise and the Bishkek Agreement between the Governments of the Republic of Kazakhstan, the Kyrgyz Republic and the Republic of Uzbekistan on joint and complex use of water and energy resources of the Naryn Syr Darya Cascade Reservoirs) demonstrated surprising similarities among transboudary water problems throughout the world.

Another interesting conclusion of these discussions involved the implications of the absence of a higher authority for the Central Asian republics to help them in resolving multi-jurisdictional disputes in the region. In the United States, conflicts among neighboring states can be resolved by the Supreme Court or at the federal level of government. In Central Asia, particularly, after the demise of the USSR, there is no clear higher authority that has jurisdiction over multi-jurisdictional disputes. However, given the fundamental differences between the situations in the USA and Central Asia over opportunities to resolve these problems at a higher level, there has been a similar lack of success in both instances in resolving these problems. So, regardless of the need to independently mediate among affected parties (as in the Bishkek Agreement) or to seek legal relief through a higher authority (like for the ACF Conflict with the US Supreme Court), there appears to be problems with both avenues of conflict resolution.

After the case study session, the pilot study participants were escorted on a half-day technical field trip of the Apalachicola River and Estuarine (i.e., Bay) by the Florida Department of Environmental Protection. The purpose of the field trip was to explain first hand the significance and fragility of the Apalachicola Bay Ecosystem to the meeting participants, and to illustrate the potential impacts of water quantity/quality degradation on the productivity of the ecosystem.

During the pilot study meeting, it was recommended:

1. To utilize an ecosystem approach for water distribution and integrated water management based on the natural needs of the environmental systems in order to facilitate long-term sustainability.

2. To develop market-based mechanisms and legal bases for effecttive management of transboundary water basins providing for sustainable development of the Central Asia.

3. To implement mutually agreed upon systems and principles for compensation and economic losses resulting from use of water resources (e.g., to create an insurance fund).

4. To create a consortium for developing rational and effective methods for the management of water resources and for effective solutions for transboundary problems.

5. To create an informational center that collects, analyzes and provides "informational exchange" among the Central Asian countries to facilitate the prevention of regional conflicts within transboundary river basins.

6. To continue the studies on transboundary water issues related to water pollution including the transport of radioactive contaminants.

7. To develop and implement a unified approach and methodologies for analysis of water quality, in compliance with the water standards of the Water Framework Directive (WFD) adopted by the European Commission and in conjunction with the ISO (International Organization for Standardization) criteria.

8. To develop monitoring systems, especially, bio-monitoring systems that can be used to understand the technical aspects of transboundary water use among affected countries.

9. To develop educational programs and training incorporating adopted curricula on issues pertaining to integrated water management systems, including model case studies.

10. To improve the overall sustainability of the ecosystems in mid- and lower streams of the Syrdarya River by improvement of effective use of the ecosystems in the area of Aydar–Arnasayak Lake basin.

11. To study and implement Central Asia model solutions for the Central Asia region based on positive practices for the solving of transboundary issues in the Apalachicola–Chattahoochee–Flint (ACF) water basin (USA.) and in Belgium. This should include: a general methodology for the establishment of river basin management plans; a conceptual approach for decision making; and defining and implementing integrated decision support.

One of the recommendations from this meeting was to organize a NATO Advanced Research Workshop (ARW) or NATO Advanced Study Institute (ASI) to discuss in-depth transboundary water problems worldwide and solutions for problems in the Central Asia region. The present NATO ARW workshop and its proceedings may be viewed as our further step in building up sustainable ecological future in the Central Asian region.

References

M. Khankhasayev, J. Moerlins, R. Herndon, and C. Teaf. Overview of NATO CCMS Pilot Study on Environmental Decision-Making for Sustainable Development in Central Asia. In Proc. Environmental Security and Sustainable Land Use, Eds. H. Vogtmann and N. Dobretsov, Springer, NATO Science Series, 2006, pp. 65–84.

Summary Report of the NATO/CCMS Pilot Study Meeting "Environmental Decision-Making for Sustainable Development in Central Asia"–Planning Meeting, 26 February–1 March 2001, Silivri–Istanbul, Turkey. NATO/CCMS. (*www.nato.int/ccms/pilot-studies/pilot005/*)

Summary Report of the NATO/CCMS Pilot Study Meeting "Environmental Decision-Making for Sustainable Development in Central Asia"–Working Group Meeting, 16–20 March 2002, The Belgian Nuclear Research Centre Headquarters (SCK-CEN), Brussels, Belgium. (*www.nato.int/ccms/pilot-studies/pilot005/*)

Summary Report of the NATO/CCMS Pilot Study Meeting "Environmental Decision-Making for Sustainable Development in Central Asia"–Working Group Meeting on Landscape Sciences, 22–24 Sept. 2002, Almaty, Kazakhstan. (*www.nato.int/ccms/pilot-studies/pilot005/*)

Summary Report of the NATO/CCMS Pilot Study Meeting "Environmental Decision-Making for Sustainable Development in Central Asia"–Working Group Meeting "Landscape Science and Public Health Issues for Environmental Decision-Making in Central Asia", 17–18 March 2003, Brussels, Belgium. (*www.nato.int/ccms/pilot-studies/pilot005/*)

Summary Report of the NATO/CCMS Pilot Study Meeting "Environmental Decision-Making for Sustainable Development in Central Asia"– Working Group Meeting on Water and Health in Central Asia, 4–5 November 2003, Almaty, Kazakhstan. (*www.nato.int/ccms/pilot-studies/pilot005/*)

R. Herndon, B. Yessekin, V. Bogachev, M. Khankhasayev, J.M. Kuperberg, J. Moerlins and I. Petrisor (editors), Environmental Decision-Making for Sustainable Development: A Central Asian Perspectives, *NATO/CCMS Pilot Study Monograph (Pre-publication)*, IICER Florida State University, March 2002.

C. Teaf, M. Khankhasayev, and B. Yessekin (editors), Risk Assessment as a Tool for Water Resources Decision-Making in Central Asia, Kluwer Academic Publishers, NATO Science Series, Dordrecht, The Netherlands, 2004.

PART II: MANAGEMENT OF TRANSBOUNDARY WATER RESOURCES IN CENTRAL ASIA AND CAUCASUS REGION

INTEGRATED MANAGEMENT OF TRANSBOUNDARY WATER RESOURCES IN THE ARAL SEA BASIN

NARIMAN KIPSHAKBAEV
*Scientific Information Center of the Interstate Commission for Water Coordination (SIC ICWC)
Almaty, Kazakhstan*

Abstract: Water management balance of the Aral Sea Basin has reached its limits, all water resources are being used, and there is no spare water available. In the future, water needs will grow. Present paper discusses the issues related to interstate cooperation in improving the water resources management system for transboundary rivers.

Keywords: Aral Sea Basin, integrated water resource management, water management, basin water organizations

1. Introduction

Water is a key factor in the social and economic development of the Central Asia countries. So far the Central Asian Region has not encountered drastic water shortages. However, economic water needs are currently met without considering environmental needs. During semi-dry and dry years, in the Aral Sea Basin, water shortage, both quantitatively and qualitatively, aggravates the economic situation and farming conditions, especially in the lower reaches of the Amudarya and Syrdarya rivers.

In the future, the water shortages will definitely become more severe, especially due to population and industrial growth, and preservation of the minimally required needs and parameters of the Amudarya and Syrdarya rivers and part of the Aral Sea.

Geographically, the boundaries of the Central Asian Region coincide with the boundaries of the Aral Sea Basin, which represents a relatively large internal water collector located in a desert and mountainous area of the Central Asia republics, and the inland Aral Sea that does not have outflows (see, Fig. 1). The Aral Sea Basin includes water collection from the rivers Amudarya, Syrdarya, Zerashvana, Turgaya, Sarysu, Chu, and Talas. The Aral Sea Basin extends approximately from 56° to 78° E longitude and from 33° to 52° N latitude, and its area is slightly over 2.2 million km².

Figure 1. Central Asian Region

2. Water Resources of Central Asia

The water resources of the Aral Sea Basin consist of surface and underground waters mainly from two rivers of the region, i.e., the Amudarya and Syrdarya. The total annual resources of the river water are 116.5 km³/year, including 79.3 km³/year from the Amudarya and 37.2 г km³/year from the Syrdarya River.

The water resources of the Aral Sea Rivers originate in several countries. For example, about 43.4% of the total flow originates in Tajikistan, 24.4% originates in Kyrgyzstan, 9.6% in Uzbekistan, and 1.2% in Turkmenistan (Table 1).

Until 1960, the water intake for all needs of the region had not exceeded 63 km³ per year, which helped preserve a stable water balance of the Aral Sea. However, increases of the population, and constantly growing needs for food for the population and raw materials for industries (e.g., cotton and rice) started adversely affecting the water balance in the Aral Sea (Table 2). It was after 1960 when intensive use of the water resources of the basin started due to rapid population growth, rapid industrial development, and, to the greatest extent, irrigation. In 1960, the total water intake in the Aral Sea Basin was 60,610,000 m^3, and, by 1990, it had increased up to 116,271,000 m^3, which is equal to the total annual water flows from the Amudarya and Syrdarya rivers. During the period of 1960 through 1990, the population in that region increased by a factor of 2.7, and the area of the irrigation land increased by a factor of 1.7.

TABLE 1. Average annual flows of the Aral Sea Basin.

State	River basin (km^3)		Aral Sea Basin	
	Syrdarya	Amudarya	km^3	%
Kazakhstan	2.43	–	2.43	2.1
Kyrgyzstan	26.85	1.60	28.45	24.4
Tajikistan	1.00	49.58	50.58	43.4
Turkmenistan	–	1.55	1.55	1.2
Uzbekistan	6.17	5.06	11.23	9.6
Afghanistan and Iran	–	21.59	21.59	18.6
China	0.75	–	0.75	0.7
TOTAL:	37.2	79.28	116.48	100

TABLE 2. Dynamics of water resources use in the Aral Sea Basin.

Year	Total, km^3/year	Including irrigation
1960	60.6	56.1
1970	94.5	86.8
1980	120.7	106.8
1990	116.3	106.4
2000	190.0	

Therefore, the current total use of natural water resources is 130–150% in the Syrdarya River Basin and 100–110% in the Amudarya River Basin. It means that the water is returned and recycled throughout the entire basin (see Table 3).

TABLE 3. Dynamics of water resources use in the republics of the Aral Sea Basin (1980).

State	Aral Sea Basin	
	km³	%
Kazakhstan	14.2	11.8
Kyrgyzstan	6.1	5.1
Tajikistan	15.5	12.9
Turkmenistan	26.0	21.7
Uzbekistan	58.3	48.5
TOTAL:	121,1	100

3. Water Resources Management

The Region is continuously improving its water resources management trying to respond to challenges of the transition period. Multipurpose use of water resources and complexity of most water management facilities require multisided satisfaction and compliance with sometimes mutually exclusive requirements of various water users, maximizing the economic effect and mitigating any undesirable environmental impacts. Under these conditions, transfer of water management authority to one of the water users causes a limited and ineffective approach to resolution of interdepartmental problems, which does happen in Kazakhstan, Uzbekistan, and Kyrgyzstan.

The following are the major international water management agreements established in Central Asia:

- 18 February 1992. Agreement among Governments of the Central Asia Countries entitled On Cooperation in Mutual Management of Use and Protection of Water Resources from Interstate Sources.

- 26 March 1993. Agreement among Heads of States of Central Asian Region entitled On Joint Activities on Resolution of the Aral Sea and Aral Region Problems, Environmental Remediation, and Social and Economic Development of the Aral Region.

- 9 April 1999. Agreement among Governments of the Central Asia Countries entitled On Status of the International Aral Sea Rescue Fund (IASRF).

- 6 October 2002. Resolution of Heads of States of Central Asian Region entitled On Major Focus Areas of the Program of Specific Actions on Improving Environmental Situation in the Aral Sea Basin for the Period of 2003–2010.

The last fourteen years of joint water use in the Amudarya and Syrdarya basins have shown that there are certain difficulties and flaws. Existing Water Management Structure is shown on Fig. 2.

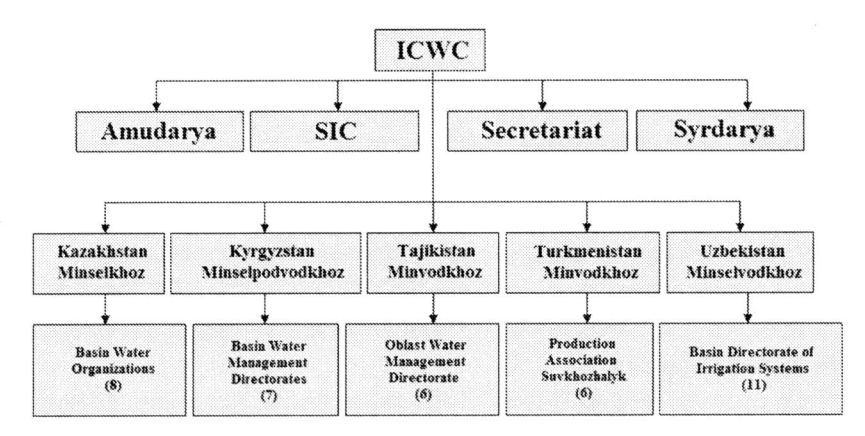

Figure 2. Existing structure of transboundary water resources management in the Aral Sea Basin

Joint management and protection of transboundary rivers is one of the major challenges. A few individuals who are responsible for developing strategies do not fully realize or even ignore close interactions between water resources management, food and energy production, industry growth, and environmental protection. All of these issues are being constantly discussed, but no visible results have been achieved yet.

- Water management agencies in Kazakhstan, Uzbekistan, and Kyrgyzstan are not authorized to regulate water use and protect water resources in their states.

- Water management agencies are not interdepartmental coordination agencies with the authority to regulate water use and protect water resources.

- Old command economy traditions are still embraced in the current executive government state and regional agencies, preventing interstate agencies (ICWC, BWO, and SIC) from implementing mutually agreed upon actions in water resources distribution among states and water users.

- Not all ICWC members are heads of the agencies authorized to regulate water use and protect water resources.

The existing water management structure will have to be revised on both regional and national levels. In this respect, water resources management in all Central Asia countries will have to be implemented by specialized state agencies on an interdepartmental basis.

On a <u>state level</u>, a state agency will have to be established with the purpose of implementing proactive scientific and technical, as well as investment policies, which will enforce rational water use and protection of water resources and resolve top priority interdepartmental and regional problems associated with the water supply for economic needs and environmental problems of the river basins.

On a <u>river basin level</u>, a Basin Water Authority will have to be created to provide comprehensive water resources management on the basin based principle.

The regional Basin Water Authority should have the following responsibilities:

- Assure stable and reliable operation of key interregional and interdepartmental water facilities.

- Enforce strict compliance with the established regulations and operational modes for operation of river trunk hydraulic structures.

- Provide environmental monitoring and compliance with all requirements on environmental and sanitary conditions of the river.

- Operate jointly with executive regional agencies responsible for water supply taking into account the needs of the ecosystems.

On a <u>water user level</u>, the reorganization should result in creation of either a consumers' cooperative group or an association of water users.

Democratization of the social and economic development of the country and requirements for smooth and conflict free environmental development require more active public participation in the water resources management process. For this purpose, the following needs to be formed:

- Coordination Council at BWO Amudarya and BWO Syrdarya, as well as at a national level agency for water resources regulation and protection.
- Basin councils at Basin Water Organizations.

In respect to this, a large scope of near-term activities on improving water resources management in the Aral Sea Basin will have to be conducted:

- Transition to integrated water resources management on both the interstate and national levels, including water quality management.
- Increased role and authority of the interstate agencies, so that they can smoothly implement their functions on distribution of water resources in accordance with the water needs and regulations established in the intergovernmental agreements.
- These agencies should be independent, to a fair degree, and they should have authority to make decisions in extreme and drastically changing water management situations.
- Rational and economical use of water resources and their protection.
- Taking into account environmental requirements, especially for the Aral Sea and Aral Sea Region.
- Transition from the water resources management to the water demand management.
- Problems associated with use of transboundary water will have to be resolved jointly, taking into account the needs and demands of all Central Asia and Aral Sea countries.
- Provide water for meeting the economic development and social needs of the countries with equality of access rights to the water supplies.
- Preserve elements of terrain and natural habitats for not only humans, but also for flora and fauna that have certain requirements for water quantity, quality, and elemental composition.
- Prevent catastrophic or emergency situations associated with water resources (floods, mudslides, draughts, etc.).

4. Conclusion

Water management balance of the Aral Sea Basin has reached its limits, all water resources are being used, and there is no spare water available. In the future, water needs will continue to grow. Future joint efforts will have to focus on interstate cooperation for improving the water resources management system for transboundary rivers, on giving more authority and responsibility to the interstate bodies, such as ICWC, International Commission on Sustainable Development, IFAS (International Fund for Saving the Aral Sea) Executive Committee and their bodies.

The top priority task for the countries of the Aral Sea Region and IFAS regional bodies is to implement the resolutions of the Heads of States of Central Asia states on Major Focus Areas of Program of Actions on Improving Environmental, Social, and Economic Situation in the Aral Sea Basin during the Period of 2003–2010. Only joint and well-coordinated activities will help to effectively resolve the water management related problems of the region.

References

Convention on Protection and use of Transboundary Water Flows and International lakes. Special Edition "International Water Flows". UN and World Bank, New York and Geneva (2000).

Proceedings of Dushanbe Meeting of Heads of States of Central Asia States on the Aral Sea problems, Dushanbe (5–6 October, 2002).

Agreement between Republic of Kazakhstan, Republic of Kyrgyzstan, Republic of Uzbekistan, Republic of Tajikistan, and Turkmenistan on Joint Management and Protection of Water Resources from Interstate Sources, Almaty. (18 February, 1992)

Dukhovny B.A. *Intergrirovannoe upravlenie vodnymi resursami na transgranichnykh vodotokakh.* (Integrated Management of Water Resources in Transboundary Flows). Course of Lectures for the ICWC Training Center. Tashkent (2002).

N. Kipshakbaev, V.I. Sokolov. *Vodnye resursy bassejna Aral'skogo moray – formirovanie, raspredelenie, vodopol'zovanie* (Water Resources of the Aral Sea Basin: Formation, Distribution, and Water Use). Proceedings of Scientific Conference on Water Resources of Central Asia. Almaty (2002).

Kipshakbaev N. *Problemy transgranichnykh vodotokov rechnykh bassejnov tsentral'noj Azii i puti ikh resheniya* (Problems of Transboundary Water Flows of the Central Asia River Basins and Their Solutions). Proceedings of the International Scientific Conference: Environmental Stability and Advanced Approaches to Water Resources Management in the Aral Sea Basin, Tashkent (5–7 May, 2003).

TRANSITION TO IWRM IN LOWLANDS OF THE AMU DARYA AND THE SYR DARYA RIVERS

VICTOR A. DUKHOVNY
Scientific Information Center of the Interstate Commission for Water Coordination, B-11, Karasu-4, Tashkent 700187, Republic of Uzbekistan

MIKHAIL G. HORST
Central Asian Research Institute of Irrigation (SANIIRI), B-11, Karasu-4, Tashkent 700187, Republic of Uzbekistan

Abstract: The integrated water resources management (IWRM) is becoming popular and gradually implementing in water sector in Central Asia. After successful completion of the IWRM–Fergana Project in the Fergana Valley area within the boundaries of three pilot systems in Kyrgyzstan, Tajikistan, and Uzbekistan, a new phase of extending the gained experience to other areas having the conditions similar to the Fergana Valley was initiated. The proposed IWRM implementation in the lowlands of the Amu Darya and the Syr Darya rivers requires taking into account socioeconomic and environmental specificities of this zone and the transboundary component of the region. Maintaining equality and sustainability of water allocation at transboundary basin level would guarantee adherence to IWRM at national level. The presentation describes the suggested IWRM mechanisms and tools for given conditions.

Keywords: sustainable development, water deficit, Central Asia, integrated water resource management, transboundary water resources

1. Introduction

Transboundary water management in the Aral Sea Basin by the five Central Asian States is a unique example in the world practice where the States plan and coordinate their joint activities, and, simultaneously, do annual planning and operative allocation of water resources of the Amu Darya and the Syr Darya rivers.

Since independence, an important step was made in the transition from the administrative "top-down" water management by the Ministry of Water Resources of the USSR to the IWRM at transboundary level. Immediately after the collapse of the Soviet Union, the Interstate Commission for Water Coordination (ICWC) was created by the five countries of Kazakhstan, Kyrgyzstan, Tajikistan, Turkmenistan, and Uzbekistan. The ICWC and its executive agencies are governed jointly by the heads of national water departments on behalf of the 5 countries. The ICWC activities, despite the increase of abnormal hydrological conditions (more frequent dry years and floods), allowed the participating countries avoid conflicts in joint river flow management and allocation.

The Central Asia Region has had a prior experience with some elements of IWRM that was based on a hydrographic unit during the Soviet period (1926–1950). It was in the systems Zerdolvodhoz, Upradik, and Kirov canal, and related to water management between provinces and between the republics, covering thousands of hectares. In the period between 1956 and 1972, IWRM was applied to the lands of Hunger Steppe during the complex development of irrigation systems, as well as in Karshi steppe, and in some other systems (1973–1990). The shortcomings of these complex approaches included an absence of the democratic aspects of management, lack of water user's participation, strong orientation the Amu Darya and the Syr Darya rivers toward state funding, unlimited water usage, and many other difficulties.

Some initial steps in changing water management were made by the water users associations (WUAs) created in Kazakhstan (from 1995 to 1999) and later in Uzbekistan (from 2000 to 2003). The WUAs took over some operational and maintenance functions for irrigation and collector-drainage network, which formerly were performed by larger cooperative farms. However, this experience cannot be considered in the gist of transitioning to the IWRM. It represented only initial elements of IWRM.

Strengthening of cooperation between the Central Asian States in transboundary water use in the lowlands of the Amu Darya and the Syr Darya rivers based on the transition to IWRM has been considered by the Interstate Commission for Water Coordination (since 1999) as urgently needed for improvement of water relationships under current socioeconomic conditions in the region.

2. Experience of IWRM Implementation in the Fergana Valley

It is clear that effective and integrated water resources management calls for both the integration at both the country and basin levels, and the interlinking, through common idea and determination, of all water hierarchical levels, and, the most important, it calls for the involvement of all national structures in the process of integration. This understanding resulted in the initiation by ICWC of the IWRM pilot project in the Fergana Valley in the territory of the three republics, Kyrgyzstan, Tajikistan, and Uzbekistan. During the four years of implementation of this project, a number of new approaches have been developed and tested.

The Fergana Valley was selected for this pilot project because a high demographic pressure in this area created complex social issues in the region due to limited irrigated area and a high density of rural population. In this region, it is very important to provide equal, equitable, and guaranteed access to water and land resources. The three countries–Kyrgyzstan, Tajikistan, and Uzbekistan–adopted a strategy for transition from socialism to market conditions; however, as our experience of IWRM implementation showed, the general principles of IWRM proved to be acceptable for all conditions.

Unlike other approaches used in many other projects, where only a part of water hierarchy is considered under the concept of IWRM, the IWRM–Fergana project covers the whole national water management framework, from the public management through basins-systems-WUA to water users.

One of the main objectives of IWRM is the creation of conditions for achieving potential water productivity by all water users, i.e., conditions for technologically feasible, cost effective, and ecologically safe level of production per unit of water withdrawn from water sources at head structures. In order to achieve this:

- Water user (e.g., farmer) should have the possibility and skills for rational water and land use, and for achieving the potential level of productivity in the field. (currently, the actual productivity is 1.5–2 times lower than this level)

- Farmers should have the possibility to sell their product at fair market prices and receive earned income.
- Water managers at all water hierarchical levels should provide equitable, stable and adequate water supplies on time, quantity and quality, and minimize unproductive water losses.

Basic principles implemented in the project:
- Management based on a hydrographic unit in order to avoid administrative interference.
- Public participation as a guarantee of "hydro-solidarity" against various types of "hydro-egoism".
- Horizontal and vertical coordination of all water users.
- Water demand management instead of water supply management.
- Priority of environmental and drinking water supply.
- Decision making on water allocation taking into account all kinds of water resources. (surface water, ground water, and return water)
- Transparency, public awareness, and accountability.
- Combination and coordination of "bottom-to-up" needs with "top-to-down" limitation.

At the present stage of the project, there have been adopted the following approaches:
- Involvement of all stakeholders in IWRM adoption and implementation.
- Creation of the institutional management framework that includes representatives of all stakeholders and supports state registration.
- Coordination of lower and upper level organizations both in the area of management and in the area of governance and tools.
- Management of the information system.
- Hydrometry.
- SCADA systems.
- Advisory system.
- Training of water users and water managers.
- System of business-plans and cost-sharing.

As a result of project implementation, water productivity parameters have been improved for all pilot systems and WUAs. At the present, the project is extending its scope.

3. IWRM Implementation in the Lowlands

Similar to the Ferghana Valley, the lowlands and delta areas of both rivers are the most socially depressed zones in Central Asia. However, the causes are different. In the Ferghana Valley, the major destabilizing factors are the intensive growth of the population and scarcity of land resources, leading to unemployment and low social living standards in rural areas. In the lowlands (except the Khorezm oasis), the land is abundant but water resources are scarce. Uneven water distribution between the upstream water formation zone and the downstream areas, especially in dry years, is a key water problem. Other deficiencies include the negligence of ecological requirements contributing to degradation of the environment of the deltas.

Therefore, the transition to IWRM in the lowlands cannot be limited (like the IWRM–Ferghana project) to a national component. There is a need for creating management systems of the entire river basins, sustainable and equitable water supply at the transboundary level and providing the opportunity for stable functioning of all elements in the water hierarchy and at the national level. This would reduce the unproductive losses at national and local levels and would increase the water productivity and, at the same time, would create the conditions for a guaranteed water supply for all water users and uses, including the environmental use (deltas, wetlands, and natural complex).

Thus, unlike to the IWRM–Ferghana project, the IWRM-lowlands project has a component related to the transboundary water management. In terms of its content, this project is similar to other on-going or implemented projects in the region, such as 'WEAMP–GEF', 'IWRM Strategic Planning' (ESCAP), 'RETA ADB'. The main elements of the IWRM–lowlands project include:

- A legal framework for guaranteed water management in the both rivers. (Agreements on rules of water management in the Amu Darya and the Syr Darya; Agreements on environmental limitations

and respective environmental needs regarding water management in the Aral Sea Basin)
- Procedures for transboundary collector-drainage water and river water quality management.
- Determination the parameters of water allocation in Amu Darya and Syr Darya downstream and midstream.
- Sustainable water supply, taking into account the needs of nature.
- Equitable and uniform water distribution at all water hierarchical levels.
- Equal and equitable water supply for all categories of water users.
- Substantial reduction of unproductive water losses.
- Efficient organizational structure of water management.
- Principles of democratic water management through involvement of representatives of all sectors, stakeholders, and primarily water users.
- Solutions of social problems related to equitability in distribution of water including drinking water.
- Improvement of water and land productivities.

4. Characteristic Problems of the Lowlands

The Amu Darya and Syr Darya downstream zones are troubled by social and environmental tension and by a certain loss of control over water resources. This became apparent during the extremely low water years 2000 and 2001.

Currently, we observe specific socioeconomic and environmental conditions caused largely by water-economic activities and consequences of irrational water management over the last 5–10 years both in the Amu Darya and the Syr Darya downstream and in the Aral Sea Basin as a whole.

It would take time and would require that IWRM be implemented at all levels of water management in addition to setting clear priorities in water resources use in order to overcome the negative trends–decline of agricultural production, environmental deterioration and manpower drain.

4.1. SOCIOECONOMICAL PROBLEMS

The main socioeconomical problems, which manifest themselves particularly in the lowlands and of the Amu Darya and Syr Darya are:

- A sharp decrease of agricultural productivity to approximately 50% (as compared with the year 1990).
- Intensified influence of the aggravated ecologic situation on the conditions of agricultural lands, fisheries, marshes, and wetlands.
- Intensified influence of the low-water years on the socioeconomic situation.

The total population in the lowlands of the Amu Darya (Khorezm province of Uzbekistan, Dashoguz province of Turkmenistan and the Republic Karakalpakstan–part of Uzbekistan) and of the Syr Darya (Kyzylorda province of Kazakhstan) is 4,845,600 people. More than 60% of that population (over 2,950,200) lives in rural areas, and agriculture is their main source of income.

The birthrate in all provinces (except Dashoguz province of Turkmenistan) decreased. This was mainly caused by the migration of young people from these areas.

The total economical losses from the migration in the Uzbek Priaralie region during the period from 1970 to 2001 was estimated as USD $20.4 million and in the Kazakh Priaralie was estimated as USD $20.65 million.[1]

The total losses due to the decline in economic activities are estimated to be USD $70 million annually for the entire Priaralie area.

Based on data collected in 1995 by the World Bank, the average income per person in this area is 1.5 to 2.5 times lower than the national average, and is often below the minimum living wage.

Recent low-water years 2000–2001 had dramatic effects on the Gross Domestic Product (GDP) towards its abrupt decrease. Before that period, the GDP in the lowlands of the Amu Darya River of Khorezm province and Karakalpakstan was increasing, and after it

---

[1] Here and further damage data was taken from the projects:
"Evaluation of socioeconomic consequences from ecological disaster–the Aral Sea shrinkage", 2001, edited by V.A. Dukhovny, Project INTAS/RFBR–1733, SIC ICWC, Tashkent.
"Economic evaluation of local and joint measures on reducing socioeconomical damage in the zone bordered upon the Aral", 2004, edited by V.A. Dukhovny. Final project report INTAS–ARAL-2000–1059, SIC ICWC, Tashkent.

started to decline. The similar picture shows GDP per capita (Fig. 1). According to national statistics, only in the Dashoguz province there is a tendency of increasing the GDP and GDP per capita.[2]

First of all, the disproportion in income is reflected in consumer food basket. Comparatively low level of food consumption in Karakalpakstan is caused not that much by shortage of food, but mainly by low income. The consumer food basket of the population of the lowlands, besides vegetables, watermelon and bread, does not correspond to physiological norms. A significant part of the population does not consume the adequate amount of protein and vitamins.

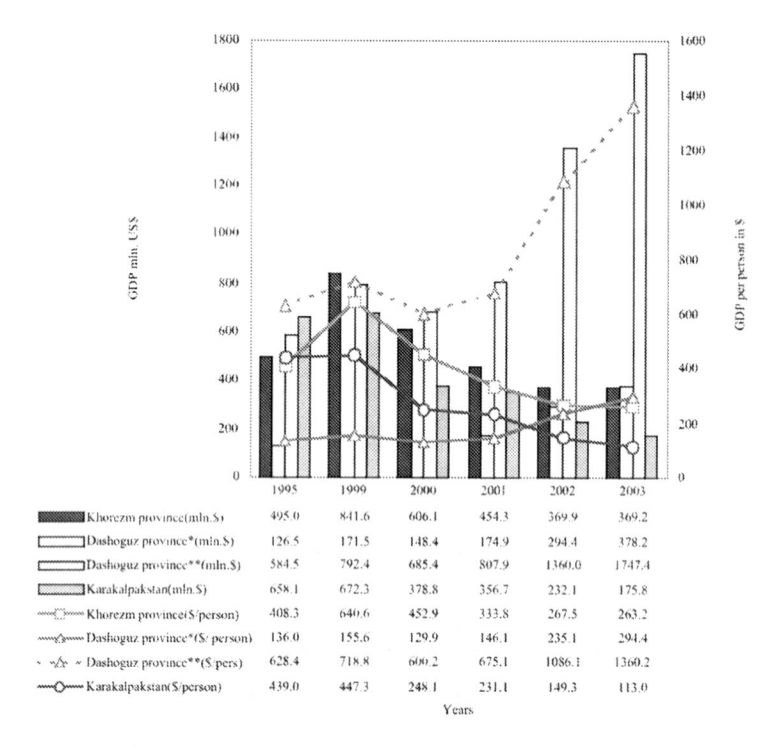

	1995	1999	2000	2001	2002	2003
Khorezm province (mln.$)	495.0	841.6	606.1	454.3	369.9	369.2
Dashoguz province* (mln.$)	126.5	171.5	148.4	174.9	294.4	378.2
Dashoguz province** (mln.$)	584.5	792.4	685.4	807.9	1360.0	1747.4
Karakalpakstan (mln.$)	658.1	672.3	378.8	356.7	232.1	175.8
Khorezm province ($/person)	408.3	640.6	452.9	333.8	267.5	263.2
Dashoguz province* ($/person)	136.0	155.6	129.9	146.1	235.1	294.4
Dashoguz province** ($/pers)	628.4	718.8	600.2	675.1	1086.1	1360.2
Karakalpakstan ($/person)	439.0	447.3	248.1	231.1	149.3	113.0

Figure 1. Dynamics of GDP and GDP per capita (*Dashoguz province estimated by the market rate of USD; **Dashoguz province estimated by the official rate of USD)

4.2. ENVIRONMENTAL PROBLEMS

Current environmental conditions in the Amu Darya and the Syr Darya downstream result from the rivers' natural volume of flow and

---

[2] Market value and official rates of the US$ vary up to 4.62 times of the value.

the human impacts, which were intensified in the upstream and midstream over the last forty years. This relates primarily to the regulation of flow by reservoirs, and also to withdrawing and/or discharging water, including drainage water and wastewater.

The current water management system in the Priaralie is based on a residual principle (water remaining after satisfying economic needs), and it is greatly aggravated by unreliable forecasts of the flow. This causes poor control of the inflow of water to the delta and its distribution, affecting discharges. Also, it results in complete dehydration of the delta, or in a sudden inflow of high water that leads to water accumulation. At the best, only from 16 to 20% of this excessive water can be utilized.[3] Such critical situations due to poor water management were observed in the Amu Darya Basin in low water period of 2000 and 2001 and in winter floods in the Syr Darya Basin in 2003 and 2004. As a result of low water levels in 2000 and 2001, the stable natural landscapes in the Priaralie have practically disappeared, and slowly degrading landscapes have started to prevail. Accordingly, the changes in the hydrological regime have led to considerable changes in the quality of river water. Increase of highly mineralized discharges and wastewater has led to substantial increases in water salinity and to deteriorated sanitary conditions of the river water.

4.2.1. Amu Darya Downstream

The environmental changes related to the decreased inflow to the delta were also reflected in the deterioration of drinking water quality due to increase in salinity and reduction of groundwater inflow.

The main negative factor in *water quality* deterioration in the Amu Darya downstream was the discharge of *return water* into the midstream. This caused the *soil salinization* process and aggravation of soil composition. The water salinity dynamics in Amu Darya downstream is shown in Fig. 2.

---

[3] "South Priaralie – new outlooks", 2003, edited by V.A. Dukhovny and Joop de Schutter, NATO Science for Pease Project, SIC ICWC, Tashkent.

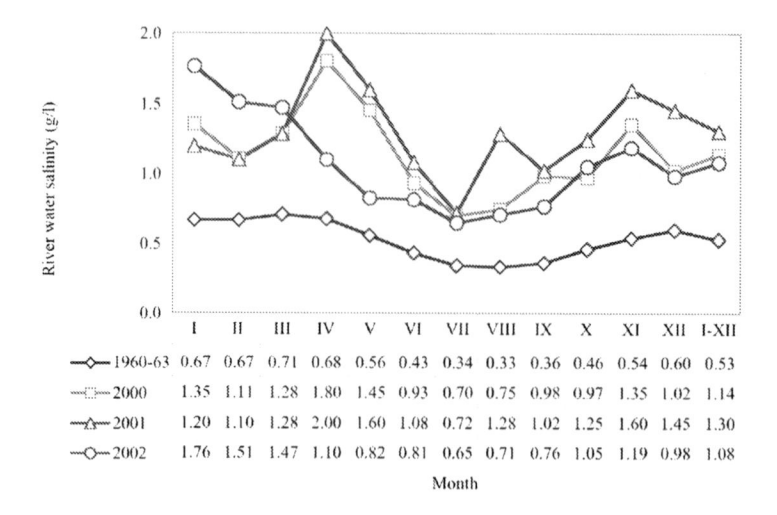

Figure 2. Water salinity (monthly and annual mean) in the Amu Darya River (Samanbai section)

In low water years of 2000 and 2001, the mean annual salinity in the Amu Darya was in the range from 1.14 to 1.30 g/l, while the mean monthly value in April was as high as from 1.8 to 2.0 g/l. The salinity has increased from 2 to 2.5 times when compared to the 1960s level of salinity.

The increase in river pollution due to the discharges of domestic sewage, industrial wastewater, and highly mineralized collector/drainage water has aggravated the environmental and socioeconomic conditions both downstream and midstream.

A similar picture can be observed in the Syr Darya midstream and downstream.

4.2.2. Syr Darya Downstream

Before the regulation of the river flow, the water salinity in the river downstream slightly varied and the changes in water availability had a little effect on the salinity values. The mineral content was from 0.6 to 0.7 g/l. Intensive irrigated agriculture in the sixties caused an increase of water salinity to 1.1 g/l during the seventies (Fig. 3).

The main cause of *water quality* deterioration in the Syr Darya downstream was a discharge of *return waters* from the Ferghana Valley and in the river midstream. Water deterioration also affected the *irrigation norms* (they increased) and the process of soil *salinization,* leading to serious degradation of lands, loss of soil fertility, reduction of crop yields and quality of agricultural production. The water quality within Syr Darya downstream *does not meet requirements of drinking water supply* and fisheries.

While the salinity upstream is no more than from 0.3 to 0.5 g/l, after the river leaves the Ferghana Valley, the salinity increases from 1.2 to 1.4 g/l. Further downstream, the water salinity reaches from 1.4 to 1.6 g/l in the Chardara section and from 1.6 to 2.0 g/l in Kyzylorda, and finally from 1.7 to 2.3 g/l in Kazalinsk.

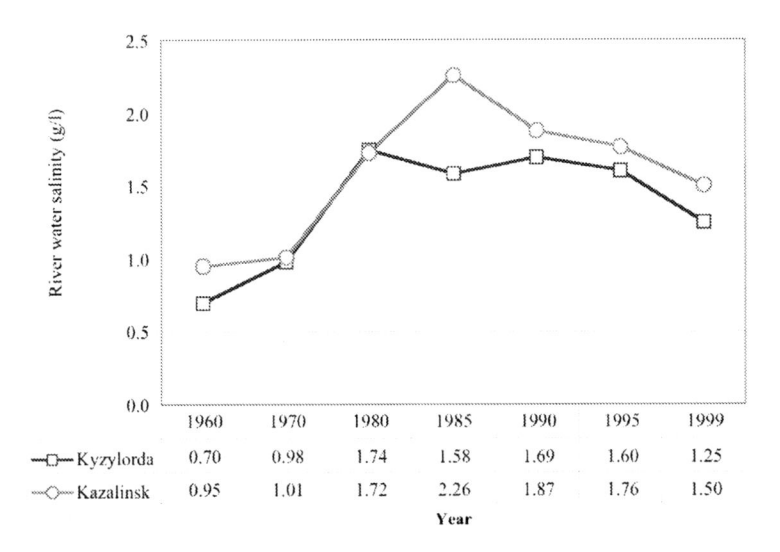

Year	1960	1970	1980	1985	1990	1995	1999
Kyzylorda	0.70	0.98	1.74	1.58	1.69	1.60	1.25
Kazalinsk	0.95	1.01	1.72	2.26	1.87	1.76	1.50

Figure 3. Water salinity (mean annual) in the Syr Darya River (Kyzylorda and Kazalinsk sections)

Unstable environmental conditions in the South and North Priaralie are aggravated by many economic and sociohygienic problems, which are associated with irrigated agriculture, unauthorized water intakes, saturation of cropping patterns by rice, and uncontrolled discharge of domestic sewage and agricultural waste water.

Moreover, during the last few years a problem related to the transboundary character of the Amu Darya and the Syr Darya occurred. It highlights the disadvantages of the lands located within the river deltas. Thus, the territories of Northern Karakalpakstan, Dashoguz province in Turkmenistan and Kyzylorda province in Kazakhstan find themselves in difficult conditions. The breach of the release schedules, water pollution and under-supply of water to habitat, nature, and national economics are typical problems.

Therefore, enormous institutional, technological and other measures are needed to prepare for the transition toward integrated water management. These measures will require that the priorities for water for ecology and drinking water supply are clearly understood.

4.3. WATER MANAGEMENT SYSTEM PROBLEMS

The key areas of water use (such as drinking water supply, irrigated agriculture, and environment) are associated with the problems that need urgent solutions. These are related to:

- Meeting the environmental water needs at the interstate and national levels.
- Improving the efficiency of water supply systems.
- Preventing an inequitable water allocation among the states irrigation systems, canals; and raising the level and stability of water supply.
- Rehabilitation of agricultural production.

  The best results would be obtained if these needs are addressed in one package in the context of each water user, irrigation system, and the downstream areas as a whole. Water conservation, the improvements of land and water productivity, as well as water quality are the focal points of the package. The particular issues to be addressed are the following:

- Unbiased and transparent evaluation of availability of water resources for various years and cycles in terms of flow (present and future).
- Integrated usage of river water, return water, and groundwater.
- Removal of the limitations regarding to technical aspects of water management.
- Compliance with the rules for water distribution at the interstate and national levels; and implementation of water-rotation schedule and reduction of organizational losses.
- Analysis of crop patterns and crop rotation.
- Adjustment of water use norms.

Lately, the efficiency of joint interstate canal management had declined. On one hand, this was caused by severe water shortages in the downstream zone (due to an improper water management within the Amu Darya River Basin as a whole), and, on the other hand, by the loss of management control and, large water losses due to efforts to separate and establish independent intakes *(hydro-egoism)* as well.

Poor water management particularly manifests itself in low-water years, especially, in the Amu Darya River.

In the growing season of 2000, the water shortage (difference between the established diversion limit and the actual diversion) increased to 11.1 km^3, which is about 30% of the limit in the Amu Darya Basin (Table 1).

TABLE 1. Water shortage distribution among riparian countries, for 2000.

Country	Shortage (km^3)	Shortage (% of limit)
Tajikistan	0.7	11
Turkmenistan	4.6	30
Uzbekistan	5.7	37
Basin as a whole	**11.0**	**30**

Water shortage distribution is related to the location within the river basin as shown in Table 2 below.

TABLE 2. Water shortage distribution between Amu Darya River reaches for 2000.

River reach	Shortage (km^3)	Shortage (% of limit)
Upstream	0.7	11
Midstream	2.7	17
Downstream	7.6	52
Basin as a whole	**11.0**	**30**

The uneven water distribution in terms of location within the river basin can be observed at the national level as well (Table 3).

The data demonstrate critical drought conditions of the lowlands and downstream of the Amu Darya during the growing season 2000.

Usually, due to non-uniform water distribution along the river and the canals, upstream water users have an advantage versus downstream users.

TABLE 3. Water shortage distribution at national level for 2000, Amu Darya Basin.

Republic, river reach, province	Shortage (km³)	Shortage (% of limit)
Turkmenistan		
Midstream	1.8	17
Dashoguz province	2.8	55
Republic as a whole within Amu Darya Basin	4.6	30
Uzbekistan		
Midstream	0.8	15
Khorezm province	1.2	36
Karakalpakstan	3.7	59
Republic as a whole within Amu Darya Basin	5.7	37

Low reliability on water forecasts and evaluations of the available water resources; lack of the data on actual stream flow and current shortages in the basin; lack of assessment for the damages caused by the latest water shortages are the key destabilizing factors leading to the loss of water management control in the basin during the growing season of 2000. These factors also had provoked the situations leading to the occurrence of the "above-limits" diversions.

Such water supply situations had almost catastrophic consequences in downstream zone. Inefficient system of control over water use, lack of the appropriate economic instruments, and legal liability elements greatly aggravated the situation.

Since 1994, the summer water releases for the Syr Darya became to a large extent dependent on the supply of electric power, fuel, and gas bartered from Kazakhstan and Uzbekistan to Kyrgyzstan. This caused the reduction of a guaranteed water supply for irrigated agriculture in midstream and downstream areas, and created shortages during the growing season and the losses of flow in winter time. More than 30 billion m³ were discharged into the Arnasai depression during the autumn/winter power-oriented releases from the Toktogul reservoir in the last decade. Due to such an operational regime, the entire natural system in the lowlands, excluding the irrigated agriculture, was under strong pressure. Water availability in the Kyzylorda province is very unstable during both dry and wet years.

The analysis of negative effects on the Amu Darya and Syr Darya rivers showed that the water shortage and flood problems need to be solved at the level of entire basin. Also, taking into account the time without breaking the natural cycle of water years into seasons and without selecting only the critical periods (growing-season phases) would be important for the analysis.

The main attention should be given to the analysis of natural and artificial shortages, resulting from the uncontrolled diversion, inaccurate evaluation of availability of water resources (including losses), inadequate management (mainly, the regulation of the reservoirs), and uncoordinated actions of the countries. In 2000, for example, the total damage in the Amu Darya downstream caused by water shortage was estimated to be USD $250 million. If water shortages during vegetation period would be distributed evenly along the entire river course, water availability in the lower river reaches could be maintained at the level of 80%. In this case the overall decrease of crop yields would be not more than 15% and the total losses of all water users from water deficiency would be not more than USD $50 million.

What does the IWRM–Lowlands Project provide for?

At the national level, the content of the project differs from the IWRM–Fergana project due to local characteristics of each of the downstream areas:

- The situation in Khorezm is similar to that of Fergana (lack of land resources, high population density), but differs in the specific conditions of soil, which are formed by deltaic stratified lacustrine sediments.
- The situation in Kzyl–Orda, Tashouz and Karakalpakstan is slightly different: abundant land resources, unstable water supply, lack of natural and artificial drainage, salinity problems, and excessive canal and collector capacities.

In order to solve these issues, special management, engineering, and reclamation approaches need to be applied to the IWRM development. However, the key directions and mechanisms should be the same as those been used in the Fergana Valley.

At the transboundary level, the IWRM–Lowlands project should provide:

- Increases in the interstate cooperation in the allocation of Amu Darya and Syr Darya flows on the basis of IWRM principles. This would be done by developing and improving of the institutional basis of existing transboundary water management organizations, such as the BWO (River Basin Water Management Organization) Syr Darya and BWO Amu Darya

- Establishment of Public Boards/Councils of the BWO including representatives of the states and provinces located within the Amu Darya and Syr Darya basins; major water users (like the hydropower stations; hydrometeorology service; the boards of management of main canals; the representatives of the deltas).
- Establishment of the division for water quality monitoring within the BWO to monitor the water quality and to perform accounting of river water. This division should prepare recommendations to the Interstate Coordination Water Commission (ICWC) and to the concerned governments for improvement of the measures regarding water quality of the watercourses, and the use of the surface, return and groundwater.
- Principle documents on transboundary flow management which are approved by the region's countries.
- Provisions of IWRM Boards/Councils, and their participation in planning and management of the Amu Darya and the Syr Darya rivers.
- Estimated values of ecological water requirements of the Amu Darya and the Syr Darya deltas.
- Assessment of available water resources in the rivers depending on the type of water year.
- Rules for regulation and distribution of water from the rivers depending on the type of water year and the specifics of the water regime.
- Rules for the BWO's activities during an extreme water year (flood, drought); regime of operation of water reservoirs including their emptying and filling.
- Financial relationships in the river flow management, and the regulation between riparian countries.

Provision about the responsibilities of the countries and individual water users for the observance of the operation regimes;

- Preparation of a set of models referring to the annual and prospective water management strategy in each river basin, taking into account the Amu Darya and the Syr Darya river interaction and the planning zones (water intakes, return water formation, and water productivity).[4]

On this basis:

- The BWO, countries, and water-using sectors would prepare their plan of activities and assess their impact on the downstream zones and neighboring states.
- The consequences of management decision and ways for reaching a consensus when making decisions could be identified.

5. Conclusions

- IWRM implementation in the lowlands of the Amu Darya and the Syr Darya rivers calls for an additional component of transboundary water management at the interstate level
- Natural and economic specificities of delta lands should be considered when implementing IWRM in the lowlands
- Involvement of all water hierarchical levels is a key factor for successful IWRM implementation at national and zonal levels.

References

Dukhovny V.A. "Irrigation and development of Hunger Steppe", Moscow, Kolos, 240 p., 1973

Dukhovny V.A. "Water-management system in irrigation zone", Moscow, Kolos, 255 p., 1984

Pereira L.S., Dukhovny V.A., Horst M.G. and others, "Irrigation management for combating desertification in the Aral Sea Basin", Tashkent, 421 p., 2005

Prefeasibility study "A move towards IWRM in lower reaches and deltas of Amu Darya and Syr Darya rivers", Report for USSD, Tashkent, 232 p., 2005

---

[4] On basis of previous activities in modeling and DSS (USAID, SIC ICWC and others)

IMPROVEMENT OF WATER RESOURCES MANAGEMENT IN THE ARAL SEA BASIN: SUBBASIN OF THE AMU DARYA RIVER IN ITS MIDDLE REACH

KURBANGELDY BALLYEV
Scientific Information Centre
Interstate Commission on Sustainable Development
International Fund for Saving the Aral Sea
15 Str. Bitarap Turkmenistan, Ashgabat 744000,
Turkmenistan

Abstract: The need to improve the water management infrastructure and, specifically, the transition from an administrative and geographical water management principle to a more integrated water management of irrigation systems, as well as creation of the Water Users' Association is discussed in the context of the National Program: Strategy for Economic, Political, and Cultural Development of Turkmenistan for the period of up to 2020.

Keywords: Turkmenistan, integrated water resource management, Amu Darya River, Lebap Province, irrigation, ground water

1. Introduction

The strategic policy for agricultural development in Turkmenistan is oriented towards independence in food production as well as providing limits on water consumption and this makes it necessary to improve integrated water resources management in the country.

The major goal of this institutional reform is to improve the effectiveness of agricultural production and provide appropriate conditions for a sustainable development of the national agro-industrial complex. The institutional reform not only makes it possible to improve national water resources management, but, also to provide opportunities for significant conservation of the water resources, reliable operations of the strategically important water supply and utilization systems.

One of the crucial aspects of the institutional reform is to provide a transition from the administrative and geographical water management principle to the integrated water management approach for irrigation systems. Implementation and schedules of this transition, as well as creation of the Water Users' Association, depends on the decisions of Turkmenistan's Cabinet Ministry. However, recently two Production Associations (Galkynyshsuvkhodzhalyk and Garashsyzlyksuvkho-dzhalyk) that used to operate on the administrative and geographical principle in Lebap Province (Lebap velayat) were eliminated, and one Production Association (PO) called PO Berzenskaya Irrigational System was established.

The major goals for the development of the program entitled: Improvement of Integrated Water Resources Management (IWRM) in the Aral Sea Basin for this subbasin are as follows:

- Evaluate the current status of the water resources management in the subbasin of the Amu Darya River in its middle reach.
- Develop a strategy for implementation of the National Program for up to 2020 for the Lebap Province.
- Develop proposals on improvement of the water resources management involving potential water users.

The Program is for the region of Lebap Province in Turkmenistan that uses surface water from the Amu Darya River within the boundaries of this province. The Kara Kum River is not included into the Program as a hydrographic unit of the river flow use. Long-term policies are based on the existing national economic planning development strategies for the periods of time of 2000–2005, 2006–2010, and 2010–2020 in compliance with the programmatic guidelines of the President and the Turkmenistan Government.

2. Issues Related to Water Resources Management

2.1. WATER RESOURCES MANAGEMENT IN THE MIDDLE REACH OF THE AMU DARYA RIVER

Amu Darya River Basin Organization (BWO Amudarya), the executive body of the Interstate Coordination Water Commission of

Central Asia (ICWC), is directly responsible for the water resources management. Its main function is to provide for an independent monitoring of the use of the Amu Darya River water.

The Middle Amu Darya Directorate of the *BWO Amu Darya* performs this function in the area of the Amu Darya River middle reach, i.e., from the gauging stations at Atamurat (Kerki) to the gauging stations at Birata (Darganata). In this function, it operates the major water intakes of the Karshinksy and Amubukharsky canals of the Republic of Uzbekistan, all water intakes in the Lebap Province in Turkmenistan, hydraulic structures in the river intakes, recycled river water return stations and key gauging posts. The gauging stations at Atamurat and Birata are important because they are capable of recording water consumption.

The gauging stations for Atamurat includes stations for Kelif, Mukry, and Atamurat located on the right bank of the river, and the station at Turkmenabat.

The gauging station for Birata includes stations for Birata and Lebap. The station for Eldgik that was eliminated earlier.

BWO Amu Darya, including its Middle Amu Darya Directorate, is responsible for preparing documentation on establishing Amu Darya river water intake limits for the states and water users for the review at the ICWC meeting, specifically:

- 10–15 days prior to the beginning of the accounting period, each state in the region submits its proposals to BWO Amu Darya identifying water intake volumes for a certain period divided into ten-days' segments for its districts and provinces.

- BWO Amu Darya and its Middle Amu Darya Directorate, together with the regional water companies and authorities, review the submitted proposals and establish water supply limits for each water user and water intake.

- The reviewed materials on the water intake for each state are submitted for review and approval at the subsequent ICWC meeting.

- After the water intake limits for the ICWC states are approved, BWO Amu Darya further update, review and approve the water intake limits for its water users for ten-days' segments of a certain period and distribute this information to its regional offices for implementation.

It should be noted that, taking into account the potential and current water resources management situation in the region, ICWC meetings approve water distribution options for all categories of water users in Turkmenistan and the Republic of Uzbekistan.

3. Integrated Water Resources Management as an Approach

The major principles of integrated water resources management are as follows:

- Provide for a stable, fair and equal distribution of water resources for the needs of the water users and for environmental needs.
- Provide for water resources and environmental management within hydrographic boundaries in accordance with morphologies of the specific basins.
- Account for all types of water (surface, groundwater, and recycled water) and climatic characteristics (precipitation and evaporation).
- Assure public participation in the management, funding, sustainability, planning, and development processes.
- Provide for close coordination of all water use agencies horizontally, i.e., among different industries, and vertically, among various hierarchical water use levels.
- Give priority to environmental needs.
- Assure water preservation and minimize water losses.
- Provide for information, assure openness and transparency of the water resources management system.
- Create a consulting service.

Integrated water resources management is a process that takes into account all existing water resources within certain hydrographic boundaries, considers the interests of various economic branches and hierarchical levels for water use, involves all motivated parties into the decision making process and facilitates effective water use for sustainability and improvements of the national welfare and environmental safety.

4. National Policy

4.1. OVERALL POLITICAL GOALS FOR THE PERIOD OF UP TO 2020

The National Program-Strategy for Economic, Political, and Cultural Development of Turkmenistan for the Period of up to 2020 defines the following three top priorities:

1. Achieve a level of developed countries; preserve economic independence and safety due to a rapid economic growth; implement new technologies; and increase labor productivity.
2. Provide for a continuous increase of the gross domestic product per capita.
3. Stimulate investment opportunities and provide for construction of a larger number of industrial sites.

The agro-industrial sector develops in accordance with the National Program: Strategy for Economic, Political, and Cultural Development of Turkmenistan for the Period of up to 2020 oriented towards meeting all public needs in food products, industrial needs in raw materials, as well as expanding export opportunities.

In the short term, special attention is given to the land reform, upgrades and reorganization of the agricultural management system, economic integration, and improvement of legal and state support for farmers.

A social class of private land owners and farmers is being formed, and favorable economic conditions for farmers of all forms of ownership were created. To stimulate agricultural production, meat, milk, and eggs prices were liberalized, farmers no longer have to sell all fresh produce to the state at regulated prices, and they are also exempt from profit tax and VAT. The government also introduced other benefits and incentives for those farmers who provide cotton and grain to the state.

The agrarian policy for the upcoming period has the following main goals:

- Provide for consistently high agricultural production growth.
- Make agriculture more efficient by boosting breeding and seed farming technologies, as well as increasing crop and cattle yields.

- Improve the structure of agriculture, making it more consumer friendly, implement advanced crop rotation technologies to improve soil fertility.
- Improve quality and increase the extent of processing of agricultural products.
- Prioritize those businesses that will help increase export potential of the country.
- Drastically improve material and technical basis.
- Make agricultural entities more specialized and site specific.

The average annual growth of gross agricultural output will be 10.1% in the period of 2005–2010.

5. Economic Milestones

The National Program: Strategy for Economic, Political, and Cultural Development of Turkmenistan for the Period of up to 2020 includes the following plans:

- Increase oil and gas extraction, for which new wells will be used.
- Construct a new chemical integrated plant for production of potassium chloride, potassium sulfate, caustic soda, and chlorine, which will increase production of mineral fertilizers by 550,000 tons per year, also, commission a new carbamide plant, which will additionally increase the production of mineral fertilizers by 350,000 tons.
- Increase the gross machinery and metal finishing products by a factor of 9 in 2011–2020.
- Provide for further development of the textile industry that will process up to 500,000 tons of cotton in the year of 2020.
- Increase wheat production by a factor of 2.9 in 2020 in comparison with 2000, raw cotton by a factor of 4.9, large horned livestock by a factor of 3.1 and fine horned livestock by a factor of 3.6.

By the year of 2020, up to 800,000 ha of land in Lebap Province is planned to become irrigated land. However, Lebap Province and Turkmenistan do not intend to request any additional water resources from transboundary water sources for these needs.

The challenge of the task is to make sure that that area will be consistently supplied with irrigation water. For this purpose, the Turkmenstan Water Management Ministry and Production Association Lebapsuvkhodzhalyk developed a program of activities for up to 2020 to provide for a sustainable development of the agricultural sector.

6. Resources and Their Use

6.1. CHARACTERISTICS OF SUBBASIN

Lebap Province is located in the east of Turkmenistan and occupies part of the modern delta of the Amu Darya River in its middle reach.

Lebap Province has an area of 93,727 km^2 (and a population of 1,160,300 people as of the end of 2001). Lebap Province contains 14 administrative districts (etraps), 4 cities and 26 small towns.

Lebap Province borders with the Republic of Uzbekistan in the northeast, with Afghanistan and Mary Province of Turkmenistan in the south, with Mary Province also in the southwest, and with Dasoguz Province of Turkmenistan in the west. Part of the border, from Birat (farganat) to Tuyamuyinsky gauging station goes along the water channel of the Amu Darya River.

6.2. LAND RESOURCES

The total area of Lebap Province is 9,372,700 hectares, including 2,270,300 hectares of reclamation lands and 1,363,000 hectares of priority use lands.

The irrigated land area is 293,600 hectares. A chief share of the irrigated land is accessible to the most cost effective gravity irrigation since it is located in the vicinity of the major water source, the Amu Darya River. However, since readily available arable lands that do not require big investments are scarce, Lebap Province tries to reclaim so called "high lands" in the Yulangyz massif, Gaurdak and Mukry massifs and Samsonovsky plateau. Development of those mountainous and remote areas requires huge investments for lifting water to higher elevations, construction of irrigation and drainage systems and preclusion of mudslides. Reclamation of sandy deserts with sand drifts and dunes also require big investments for construction, claying, and melioration, and also for installation of irrigation systems in light soils that require a special "closed circuit" design to reduce irrigation water losses and additional loads to the drainage system.

According to the overall estimates, 65,095 hectares of reclaimed land (22.1%) are in a good condition, 134,667 hectares (45.9%) are in a satisfactory condition, and 93,806 hectares (32%) are in an unsatisfactory condition where up to 50% of the crop is regularly lost.

The quality of the reclaimed lands in Lebap Province is also aggravated by a poor technical condition of the irrigation drainage and collector systems, drainage water runoffs without use, continuous decrease of groundwater levels and soil salination.

In addition, Lebap Province has collectors of the Republic of Uzbekistain: the Bukharsky Collector with the annual flow of up to 600 million m^3 and the Southern Collector with the annual flow of up to one billion m^3 that discharge collector and drainage water (CDW) to the Amu Darya River.

The major agricultural crops are as follows:
- Cotton   – 112,649 ha
- Winter grains – 163,646 ha
- Other crops   – 17,212 ha.

The President of Turkmenistan identified tasks on expanding irrigated lands in the National Program: Strategy for Economic, Political, and Cultural Development of Turkmenistan for the Period of up to 2020 that was approved and accepted on 15 August 2003. According to this Program, the area of irrigated lands in the Lebap Province will be increased up to 800,000 hectares by the year of 2020.

The intention is that, under any option of the economic development of Turkmenistan, none of the provinces of Turkmenistan, including Lebap Province, will request any additional water resources. Therefore, in all spheres of water use, it is recommended that advanced water conservation technologies be implemented, rational water use be enforced, water losses be minimized, irrigated land structure be optimized, and crop rotation be commonly implemented and enforced to assure high yields and maximum water irrigation benefits.

6.3. WATER RESOURCES

6.3.1. Surface Water

The major source of surface water is the Amu Darya River. Within Turkmenistan, it goes from the Kelif gauging station to the Tuya–Muyunsky reservoir. The length of the Amu Darya River in Turkmenistan is 744 km. The major user of the surface water is agriculture which consumes up to 98% of its total volume.

Utilities use 1.1% of the surface water resources, industry 1.06%, and other uses 0.04%. The Amu Darya River water is not used for electrical power production.

Mud floods are frequent in mountainous regions of Lebap Province due to long-lasting rains, washing off scarce fertile soils and causing significant economic losses.

6.3.2. Groundwater

The State Natural Resources Commission recorded 855,700 m^3/day of groundwater from 12 sources, and the predicted supply is 2,995,300 m^3/day in Lebap Province. Most settlements in the Province have sufficient quantities of the approved supplies. Many lenticular veins were found in Lebap Province that could be used for water supply in local rural settlements. The depth to groundwater ranges from 0.5 to 25 m.

Ground water replenishes mainly from infiltration of the Amu Darya river water, water from the irrigation channels and irrigated lands.

6.3.3. Water Quality

Not all gauging stations have the capacity to measure water quality. Only the gauging stations at Atamurad and Darganata on the Amydarya River have this capability.

For a number of years, Turkmenistan research organizations have performed a wide range of studies to identify the irrigation needs and required lengths of the drainage lines, depending on the mineral content of the water used for irrigation that can be used for economic evaluations of mineralized water use for irrigation purposes. For example, the irrigation norm for the water mineralization of 1 gram/L for poorly pervious soils (medium loam with poorly permeable layers) increases by a factor of 1.11, for 1.5 gram/L – by a factor of 1.19, for 2.0 gram/L – by a factor of 1.29, etc.

Increased needs for irrigation water call for increased throughput capacities of the irrigation canals and lengths of the drainage lines, which require larger capital investments and higher operational costs. For example, if we have weak permeable soils with the water mineral content of 2.0 g/L, the drainage line will be 1.28 times longer than under the normal conditions.

According to the scientific evaluations, crop losses in the soils with weak salinity may reach 10–15%, with medium salinity 25–30%, and with heavy salinity 50–60%.

7. Water Users

The annual water intake for Lebap Province ranges from 3.5 to 5.5 billion m^3. The water intake to Karshinsky backbone canal is 4.9 billion m^3, including Turkmenistan's share, which is 0.18 billion m^3. The Amubukharsky backbone canal takes 4.6 billion m^3, and 0.3 billion m^3 of this goes to irrigation of the Farapsky Etrap lands.

The major use of this water is for agriculture which consumes up to 98% of the total. Utilities use up to 1.1% of the water resources, various industries use up to 1.06%, and up to 0.04% is utilized for other needs. The Amu Darya River water resources are not used for production of electric power.

8. Power Engineering

Turkmenistan does not use its water resources for the production of electric power. Maryjskaya hydropower plant uses water for cooling the turbines and returns it to common use. Turkmenistan does not plan to build any major hydropower plants in the foreseeable future.

9. Industry

According to the Strategy for Economic, Political, and Cultural Development of Turkmenistan for the period of up to 2020, Lebap Province plans to build a textile integrated plant, cotton spinning mills, clothes factory, cement plant in Magdanly, plants for manufacturing construction materials (brick). These facilities will not require a significant increase in water resources needs. Therefore the economic plans of industrial development in the region will not cause a significant increase in water consumption. The forecast for water needs of the Lebap province is reported as the following:

- 2005 – 60 million cubicmeters
- 2010 – 70 million cubicmeters
- 2015 – 82 million cubicmeters
- 2020 – 123 million cubicmeters

Major problems with water use for all economic needs of Lebap Province occur during dry years when the water consumption allowable limits are decreased. During dry years, the following measures are enforced:

- Prioritize all economic activities depending on their critical water needs.
- Introduce a mandatory water consumption minimization policy at all industrial sites.

As recommendations, the following should be noted:

- Build a waste water treatment plant at those sites that use water to assure water recycling and reuse for industrial needs.
- Assure complete accounting of water use.
- Enforce fines and penalties for those individuals who violate water use regulations.

10. Environment

The environment is not considered a separate and individual water user. The assumption is that the water losses from open water sources associated with current economic development, and the water sources themselves, to a certain extent, sustain the environment, i.e., the flora and fauna that have existed for centuries, and there may be no need to interfere with this process.

Two factors should be noted regarding protection of the environment:

- Strictly enforce requirements on sanitary outflow volumes throughout the entire river length to assure sustainability of the ecosystem of the river.
- Preclude the discharge of collector and drainage water into the river. According to preliminary estimations, the annual damage to Turkmenistan economics resulting from such discharges is about 1 trillion manats.

The final factor is associated with huge expenses, the bulk of which are covered by Turkmenistan that started implementing the project on the Turkmenistan Lake of the Golden Age.

11. Urban and Rural Water Supply

The Turkmenistan Government focuses its attention on providing a reliable supply of a high quality drinking water to the population. The water supply issues are among those key issues that are directly related to further improvement of the country's quality of life. These issues include, but are not limited, to providing free gas, electricity, table salt, and water for household needs.

Water for public needs is taken, for the most part, from open surface sources, and, to some extent, recently underground water of an appropriate quality became used for household needs. The existing data show that 79.5% of the rural population of Lebap Province has a fairly high quality drinking water. In the rural areas, this number still remains relatively low, i.e., 36.3% (as of 1999). For household needs of the urban and rural population of Lebap province, only 86.8 million m^3 of water was used in 2000, including 3.4 million cubicmeter of groundwater.

To resolve problems related to providing a high quality drinking water to the Turkmenistan population, the President of Turkmenistan issued Resolution #1690 on 10 March 1994: On Expanding Capabilities of Using Underground Water for Water Supply Needs of the Turkmenistan Population and Resolution #2341 on 1 September 1995: On Improving Water Supply for Balkansky and Lebap Provinces.

As a follow-up of Resolution #1690, in 1994, the Turkmengiprovodkhoz Institute developed a Long-Term Concept for use of underground fresh water for household needs of Turkmenistan population that describes problems and their potential solutions for both

Turkmenistan and, in particular, for Lebap Province up to the year of 2010. All above mentioned documents focus on resolving the issues of providing high quality drinking water for the Lebap province population using relatively fresh underground water (salt content of up to 1.5 g/L).

12. Problems Analysis

The Problems Analysis Section of the Program describes four factors that affect the sustainable development of the water management system and, ultimately, effective continuous development of the agricultural sector:

- Ameliorated condition of the irrigated lands.
- Status and effectiveness of irrigation and drainage infrastructure.
- Operation and management of the irrigation systems.
- Institutional structure.

In the Lebap Province, the area of irrigated lands in good condition is estimated to be equal to 74,422 ha (25.6%), in satisfactory condition – 127, 695 ha (43.9%), and in unsatisfactory condition – 88,669 ha (30.5%). It should be noted that, for the land in unsatisfactory condition, 72,460 ha are unsatisfactory due to unallowable groundwater levels, 8,200 ha are unsatisfactory due to the very high salinity of soils, and over 8,000 ha are affected by a combination of these two factors.

The top priority measures to be taken in Lebap Province to improve amelioration of the irrigated lands are as follows:

- Refurbish collector and drainage systems in the area of 34,786 ha.
- Perform operational flushing in the area of 54,242 ha.
- Construct collector and drainage systems in the area of 34,054 ha.
- Capital flushing in the area of 8,200 ha.
- Refurbish collector and drainage systems and perform operational flushing in the area of 30,983 ha.
- Construct collector and drainage systems and perform their operational flushing in the area of 49,869 ha.

- Repair collector and drainage systems and capital flushing in the area of 4,980 ha.

The existing water management system has been functioning fairly well until recently, however, a transition to the market economy and an urgent need for water use minimization to meet growing agricultural needs make it necessary to upgrade the water management system.

The most optimal approach is to perform transition from the administrative principle of water management to the irrigation system based principle.

13. Hydrological Network, Data Monitoring and Management

The middle reach of the Amu Darya River starts from the Kelif gauging station and ends in the Tuya–Muyansky Water Reservoir (the length of the middle reach is 744 km).

In this area, Turkmengidromet monitors water level fluctuations and measures water consumption. Lebap Province has two hydrological stations, Atamurat (Kerkit) and Birata (Darganata).

Atamurat station consists of gauging posts Kelif, Mukrym, and Atamurat that are located on the right bank of the Amu Darya River.

Birata station consists of gauging stations at Birata, Lebap, and Il'chik that are located on the right bank of the Amu Darya River. Atamurat and Birata gauging stations measure both water level and water consumption.

All hydrological stations for water level observations and water use measurements are equipped with appropriate water accounting instrumentation, vehicles, boats, and human resources.

14. Strategy for Implementation of the National Program

14.1. DESCRIPTION OF ACTIVITIES REQUIRED
FOR IMPLEMENTATION OF THE POLICY GOALS

The major steps for achieving the established goals on improving the water management system are as follows:

- Update the existing and develop new regulatory requirements oriented at rational use of land and water resources
- Improve administrative control on all levels of use and management of land and water resources
- Implement technological upgrades of the water management system infrastructure
- Optimize economic relations between water suppliers and users.

It should be noted that measures such as updating the existing and developing newly regulations, improvement of administrative control and economic relations do not require investments and can be effectively implemented in the near future. However, implementation of technological infrastructural upgrades in the water management system require significant capital investments that can only become effective if three other above mentioned conditions are met.

14.2. PROPOSED MEASURES ON UPGRADING WATER MANAGEMENT INFRASTRUCTURE

Strategies and policies for sustainability of water resources and long-term water needs are based on the Government Programs of social and economic development of the country. The major goal of these programs is to provide sufficient amount of major food items to the population. In this respect, Turkmenistan developed a number of activities on increasing capacities of the existing water reservoirs and building new reservoirs, refurbishing the irrigated lands and making the efficiency rate of the irrigation systems 0.67 by the year of 2010 and 0.75 by the year of 2025.

Turkmenistan and Lebap Province, in particular, will gradually decrease water consumption when the above mentioned activities are implemented. However, due to the population growth and its growing needs in major food items, the area of irrigated lands will have to be increased, and irrigation water for those newly irrigated lands will have to come from implementing water conservation measures. Therefore, even with the specific water consumption going down, the overall water consumption will increase due to use of waters with a low salt content.

14.3. NEEDS IN INVESTMENTS FOR IMPLEMENTATION OF PROPOSED MEASURES FOR UPGRADES OF THE WATER MANAGEMENT INFRASTRUCTURE

The major cost elements that require potential investments are as follows:

- Capital investments for new land development.
- Capital investments for comprehensive refurbishment of irrigated lands (CRIL).
- Capital investments for construction and repairs of collector and drainage systems in the lands that do not require CRIL.

Turkmenistan Minvodkhoz identified specific capital investments for the above mentioned activities for the period of 2005–2010:

- For new land development: 9.7 million manta/ha.
- For comprehensive refurbishment of irrigated lands: 5.0 million manta/ha
- For construction and repairs of collector and drainage systems in the lands that do not require CRIL: 4.4 million manta /ha.

The state bear all costs associated with capital investments to irrigated farming (new land development, comprehensive refurbishment, construction of new water management facilities, etc.)

The state covers costs on operation and maintenance of interdepartmental irrigation systems. Operation and maintenance of intradepartmental irrigation systems are covered by farmers (3% allocations from the cost of the produce grown on the irrigated land). The existing funding mechanisms may be revised when an irrigation water delivery fee is introduced and the Association of Water Users and Farmers is created.

15. Summary and Conclusions

The major stimulating factor is the creation of economic incentives for farmers in rational use of land and water resources. In Turkmenistan water for irrigated farming is supplied for free within the allowable limits. There is a system of fines (threefold payment of the minimum charge) if more than allowable amount is taken. In addition, there are penalties for the following:

- Illegal discharge of irrigated water.
- Violation of approved watering methods.
- Unauthorized water use.
- Violation of primary water accounting requirements, etc.

Economic methods appear to be the most effective methods for assuring water conservation. And, for this purpose, two principles are to be implemented:

- Rewards for rational water use and water conservation.
- Penalties for careless water use.

References

Durdyev A.M., Problems of Climate Change and Sustainable Development, Ashkhabad, 2002.
Zavyalova L.N., Preparation for Small Scale Projects on MChR, Ashkhabad, 2005.
Berkovsky B.M., Kuzminkov V.A., Renewable Sources of Energy for Human Needs, Nauka, M., 1987.
Bezrukikh P.P., Economic Problems of Non-Traditional Power Engineering Energiay: Econ. tekhn. Ecol., 8, 1995.
Pendzhiev A.M., Potential use of Renewable Sources of Energy in Turkmenistan. Journal Problems of Desert Development, 2, 2005.
L.E. Rybakova, Use of Solar Energy, Ashabat, Ylym, 1985.
Lozanovskya I.N., Orlov D.S., Sadovnikova L.K., Ecology and Protection of Biosphere from Chemical Contamination, Nuka,Moscow, 1998.
Dyrdyev A.M., Pendzhiev A.M., Reduction of Energy Antropogenic Load to Turkmenistan Climatic System Using Non-Traditional Energy Sources. World Climate Change Conference, Moscow, pp. 319–320, 2003.
Smirnov B.M., Earth,Atmosphere and Energy, Moscow, Znanie, pp. 64, 1969.

INTEGRATED WATER RESOURCE MANAGEMENT OF TRANSBOUNDARY CHU AND TALAS RIVER BASINS

ELENE RODINA, ANNA MASYUTENKO, AND SERGEI KRIVORUCHKO
Department of Environment and Sustainable Development, Kyrgyz–Russian Slavic University, Bishkek, Kyrgyzstan

Abstract: The article includes results and outcomes obtained during the implementation of the EC/TACIS "ASREWAM Aral Sea 30560" Project. The overall objective of this project was to analyze the needs and to improve the potential of water management organizations in the field of integrated approach, planning and management of the water resources of transboundary Chu and Talas rivers. This paper is focused on the assessment of ecological and environmental aspects of the process of integrated water resources management.

Keywords: Integrated Water Resources Management, Chu and Talas rivers, EC/TACIS ASREWAM Aral Sea Project

The definition of Integrated Water Resources Management (IWRM) was given by the Global Water Partnership in 2000 as a process which facilitates the development and management of water, land and their interrelated resources in order to achieve maximum economic and social benefit in an equitable manner without undermining the sustainability of natural ecosystems. This definition strongly correlates with the concept of sustainable development adopted by the UN Commission for Sustainable Development (CSD) presented schematically in Fig. 1.

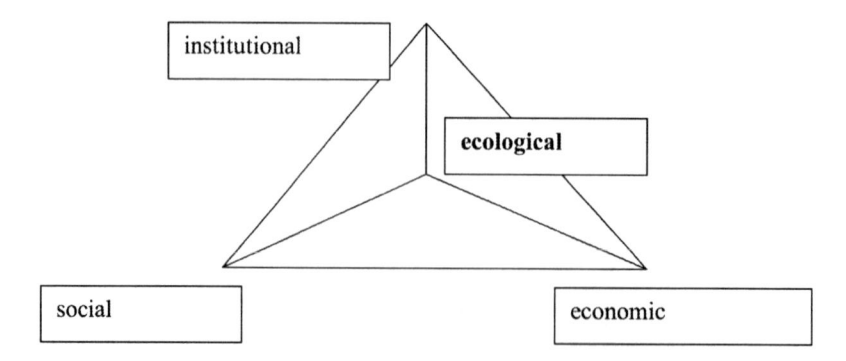

Figure 1. Four dimensions of sustainability of the UN CSD, 1995 (institutional, ecological, social, and economic)

Figure 1 shows that sustainable development, similarly to IWRM, assumes a balance of interests within the process of ecological and socioeconomic development. According to the world experience this balance can be achieved only under the condition of establishing a coordinating political management structure. In the considered case of the water resources management of transboundary Chu and Talas rivers, such a structure should be represented by the Joint Kyrgyz–Kazakh Commission on Management of Transboundary Water Waterworks Facilities on Chu and Talas rivers. The process of establishing of this Commission was going in parallel to implementation of the ADB ASREWAM Project. This process was somewhat delayed since the Commission has been established only in the middle of 2006.

We understood that at early stages the newly established commissions would focus on the issues related to water apportioning, and only after this stage we would witness the desired gradual transition to integrated water resources management. To provide this transition with a scientific and methodological basis adapted to local circumstances, a Strategic Memorandum was prepared in the framework of the ASREWAM Project. This Memorandum includes analysis of all aspects (institutional, ecological, and socioeconomic) related to the IWRM in the area transboundary river basins.

In the process of preparing the Memorandum, the concept of ecological space for ambient environment which can be used to measuring sustainability of development or sustainability of the IWRM was taken into account. (see, Fig. 2).

Overconsumption: Ecologically unsustainable area
Sustainable mode of life
Poverty: Socially unsustainable area

Figure 2. Concept of ecological space for ambient environment

The concept of ecological space for an ambient environment allows for the identification of clearly-set targets during the use of specific kinds of natural resources, including water. There is an underlying conceptual understanding that each and every activity, action, or any strategic development program aimed at both poverty alleviation and natural ecosystems protection would lead the achievement of sustainable development.

The lessons we have learned during the ASREWAM Project implementation need to be considered by the Interstate Commission. These lessons are the following:

LESSON 1. HIGH LEVEL OF MUTUAL DISTRUST AMONG EXPERT GROUPS OF TWO COUNTRIES AT THE INITIAL STAGE

During the Soviet period the problem of water apportionment in Central Asia was regulated on a centralized basis. The water for energy production was discharged from the waterworks in winter seasons, while for irrigation – in summer season. This was the essence of the water apportioning conflict, and the compromise was found in the following manner: It was decided to accumulate water for irrigation in the first hand, while water for winter discharge to produce energy was substituted by in-kind energy resources in the form of coal, oil, and natural gas supplies. Since the independence, these in-kind energy resources supply were initially curtailed, and then stopped. The experts' groups found themselves being in opposition to each other by trying to acquire as much benefit as possible for their countries. This could be explained by appearance on the stage of new groups of stakeholders such as municipal and industrial water services experts, ground water experts, socioeconomic and environmental experts, NGO, business, local self-management authorities, etc.

An analysis of this situation identified the following main causes:

1. There is no possibility to conduct full, transparent and timely assessment of water resources due to severe deterioration and, in some cases, simple destruction of water measuring outlets and facilities located on Chu and Talas rivers.
2. According to the opinion of Kyrgyz representatives, the Intergovernmental Agreement signed during the time of former USSR (1983), which governs the process of water apportioning in the area, does not fully reflect the distribution of additional water resources during vegetation period on the territory of Kyrgyzstan
3. The snow blanket measurements in Kyrgyzstan have not been done because it requires significant funding that is currently not available in Kyrgyzstan. This information is valuable and needed for Kazakhstan for making proper water supply forecasts. The Kyrgyz–hydromet was not able to conduct a correct estimation of the precipitation volumes in April and during the vegetation season. It caused frequent errors in officially provided forecasts.
4. Absence of the actual water volumes used during vegetation period.

The estimations performed within the framework of the ASREWAM Project confirmed that the cumulative volume of water resources available for interstate distribution is significantly less than that established by the 1983 Agreement and the existing water management plans. The volumes of water apportioned in the course of one year could differ from those stipulated in the Agreement and could serve as the reason for certain distrust and discords, especially, in the absence of a regular monitoring of the lateral inflows (collected from the water measuring outlets) conducted on a daily (or at least ten-day intervals) basis.

In order to remove these sources of distrust, the ASREWAM Project recommended computerizing the water runoff process to provide distribution of data and to speed-up data processing on the basis of automated monitoring and Internet-based data transfer.

To start this process ASREWAM was tasked to develop:

- A database for the Talas basin water management authority (WMA) which allows it to compare all historical data using improved approach and to regularly register ten-day interval information. The database will automatically check incoming data to identify explicit errors and prepare reports which are normally used by WMAs. The ASREWAM Project has arranged a training seminar for Kazakh and Kyrgyz WMA specialists.

- Software for seasonal planning of water resource management based on the methods of water balance compilation which in turn are used for the preparation of seasonal plans to manage and monitor these plans during the season. The software directly sends direct inquiries to the database without the need of data reentering. It also compiles standard reports used by Water Departments of Kazakhstan and Kyrgyzstan. The software issues warnings about critical water thresholds and recommendations on when the water level of Kirov reservoir (Talas River) should be dropped in order to avoid inundation.

However, this could be seen as only a beginning of an effective process of water distribution and apportionment. Among the next tasks are the improvement of the forecast system; reconstruction of hydro-meteorological infrastructure; and many others.

In order to ensure a civilized involvement of other stakeholders in the process of water consumption and public awareness on the issues of water apportioning, a Project under the aegis of Milliekontakt (the Netherlands) has been initiated in Kyrgyzstan. This project is aimed at participation of public and self-management authorities in organizing water distribution at local level. Two pilot settlements (located one by one downstream) in each of seven provinces have been selected. This will allow for in analyzing of transboundary aspects of water resources in terms of separate countries The Projects involves 18 NGOs and self-management authorities of all 14 pilot settlements (villages).

LESSON 2. WEAK INTERSECTORAL INTERACTION DURING PREPARATION OF TERRITORIAL DEVELOPMENT PROGRAMMES OF CHU AND TALAS RIVER BASIN

Since gaining independence, there was a considerable decrease in the use of arable lands in Kazakhstan and Kyrgyzstan, and the issue of water resources deficit was diminished as well. However, the process of putting agricultural lands back into use has already resulted in an increase for the demand for water resources. For instance, development plans of the Djambul province of Kazakhstan include as a target to double arable lands by adding 80,000 ha of land. To achieve this goal it would be needed to use 100% of the water runoff on Talas

River. What would be left for Kyrgyzstan if presently the Talas River runoff is divided proportionally, i.e., 50% for each country? The development plans for Talas province of Kyrgyzstan do not consider the water resources demand for municipal use in connection to the potential of the river and underground runoff. And there are many other examples.

LESSON 3. DESPITE ACHIEVED PROGRESS IN FIELD OF WATER AND ENERGY PARTNERSHIP, POTENTIAL FOR TRANSBOUNDARY COOPERATION ON ALL IWRM ASPECTS IS STILL WEAK

There is still nothing in the Kyrgyz agenda to accede the Helsinki Convention on transboundary water resources. The main barrier, for instance, in Kyrgyzstan, is the lack of accord of terminology adopted in the Convention with the actual water management practice prevalent in the country. The process of matching of official terminology is currently conducted by the Department of Water Resources and other relevant institutions, though this is evidently a long-lasting process.

However, Kyrgyzstan, similarly to Kazakhstan, has acceded to the Convention on Environmental Impact Assessment (EIA) in transboundary context. In order to work out the EIA procedures, the specialists from the State Agency of Ecology and Forestry of the Kyrgyz Republic in cooperation with NGOs selected as a pilot object the gold deposit located upstream of the Talas River, where the construction of a gold-mining facility has been started. At present, the process is going on, however it is meeting difficulties. The fact that the project is supported by the European Union brings about a certain degree of confidence that the EIA procedures will be worked out and will accelerate the adoption of the Helsinki Convention on transboundary water resources by Kyrgyzstan.

LESSON 4. ECOLOGICAL ASPECTS ARE STILL MISSING IN THE WATER RESOURCE MANAGEMENT SYSTEM AT NATIONAL AND TRANSBOUNDARY LEVELS

Evaluation of interrelations between the environment and a human activity supports the fact that environmental degradation increases its influence on the quality of life: people's health, poverty level, economic development, and even national security. For instance, pollution of riverine reservoirs results in a decreased productivity of irrigated areas and, as a consequence, to further decline of living standards of rural population.

In 2001, Kyrgyzstan received a note from Kazakhstan MFA warning to avoid the pollution of transboundary rivers Talas, Chu, and Yassa. If no serious measures are taken, this conflict may lead to the development of security issues at national level.

Our estimations for the Talas River indicate that 30–35% of the river flow rate is required in order to dilute a polluted cumulative discharge coming from the Talas municipality. The process of self-purification of discharged polluted water ends up within the borders of Kyrgyzstan. Therefore, the irrigation within Kyrgyz territory consumes only 15–20%, i.e. less than 50% of the total river flow assigned by the Interstate Water Apportioning.

All lessons learned from this analysis are presented in the Strategic Memorandum on IWRM of Chu and Talas rivers which should be used by the Joint Kyrgyz–Kazakh Commission on Management of Transboundary Water Waterworks Facilities on Chu and Talas rivers.

CHU–TALAS ACTIVITIES

LEA BURE
Economic & Environmental Officer
OSCE Centre in Almaty, Kazakhstan

Abstract: In January 2000, the Governments of the Republic of Kazakhstan and the Kyrgyz Republic signed the Agreement on Utilization of the Water Facilities of Interstate Use on the Chu and Talas rivers that are shared by both countries. Under the agreement, Kazakhstan has an obligation to reimburse a part of Kyrgyzstan's expenses for operation, maintenance, and rehabilitation of a number of dams and reservoirs located on the territory of Kyrgyzstan, but supplying water to Kazakhstan. The Agreement was ratified by parliaments of both countries and came into force in February 2002. With the involvement of the OSCE and other entities a project to implement this agreement was initiated. The project consisted of four main components. The first was to develop the statutes and regulations for the Bilateral Commission on the Chu and Talas rivers. The second component called for allocation of costs for operation, management, and rehabilitation of selected water control projects, or the outlining of the financial viability of the Commission. The third component of the project involved public participation. The final component consisted of a public awareness campaign. With support from the major donors, Great Britain and Sweden, and with additional support from Estonia, OSCE was able to realize significant successes in the Chu–Talas project, even though no comparable cooperative project exists in Central Asia. This project has demonstrated that cooperation can function, and that cooperative agreements of the sort pioneered in this project can and do have significant benefits for all participants.

Keywords: Chu and Talas rivers, Bilateral Commission, public participation

1. Basic Information about OSCE Scope of Work

The OSCE is primarily a security organization. As part of its comprehensive approach to security, it is concerned with economic and environmental issues, operating on the premise that economic and environmental solidarity and cooperation can contribute to peace, prosperity and stability. Conversely, economic and environmental problems that are not effectively addressed can contribute to increasing tensions within or among States.

In the Helsinki Final Act (1975), the participating States divided the CSCE's (now OSCE) areas of activity into three dimensions or baskets.

- Political Dimension.
- Economic and Environmental Dimension.
- Human Rights Dimension.

The second dimension is the Economic and Environmental Dimension, dealing with issues such as economic development, science, technology, and environmental protection in their relation to international security.

The tradition of the OSCE and, indeed, the mandate of the organization speak to the importance of international cooperation and mitigation of potential conflicts. This has been identified and acknowledged as an area wherein the OSCE as a competitive advantage, and the effectiveness of the OSCE's facilitation of international cooperation is clearly manifested in the Kazakhstan–Kyrgyzstan transborder project entitled, "Creation of a commission between Kazakhstan and Kyrgyzstan on the Chu and Talas rivers", which was conducted collaboratively with the Kazakhstani and Kyrgyz Governments, with the invaluable support of the United Nations Economic Commission for Europe (UNECE), the United Nations Economic and Social Commission for Asia and the Pacific (UNESCAP) and the Asian Development Bank (ADB). The project was initiated following the UN Special Program for the Economies of Central Asia (SPECA) began calculated guidelines for water management, and it was determined that a more practical and concrete mechanism would be required.

2. A Bit of Background Information

In January 2000, the Governments of the Republic of Kazakhstan and the Kyrgyz Republic signed the Agreement on Utilization of the Water Facilities of Interstate use on the Chu and Talas rivers that are shared by both countries. Under the agreement, Kazakhstan has an obligation to reimburse a part of Kyrgyzstan's expenses for operation, maintenance, and rehabilitation of a number of dams and reservoirs located on the territory of Kyrgyzstan, but supplying water to Kazakhstan.

The Agreement was ratified by parliaments of both countries and came into force in February 2002. The countries have already developed preliminary arrangements for sharing costs of exploitation and maintenance of the water management infrastructure. But according to the Agreement, a permanent bilateral commission is to be established in order to operate the water facilities of interstate use and define and share the costs for their exploitation and maintenance.

A request was made by the Governments of Kazakhstan and Kyrgyzstan for assistance in establishing the commission stipulated in their Agreement, in order to ensure effective implementation of the Agreement on the Chu and Talas rivers. UNECE, UNESCAP in cooperation with OSCE decided to joined forces and with the financial assistance of the Governments of UK, Sweden, and Estonia, began in 2003 the implementation of the Creation of a commission between Kazakhstan and Kyrgyzstan on the Chu and Talas rivers.

To begin by offering a limited description of the project: The Chu and Talas rivers are present in both Kazakhstan and Kyrgyzstan, and the separation of these two states after the dissolution of the Soviet Union has caused significant challenges from a water management perspective. Kyrgyzstan, with its mountainous terrain, has an abundance of water flowing from mountain glaciers. Kazakhstan, however, currently has a stronger industrial and urban infrastructure. The division of these states has left Kazakhstan with a chromic shortage of water, particularly in the southern areas, while Kyrgyzstan faces a shortage of resources to maintain the water management infrastructure. The potential for cooperation is evident, although the technicalities associated with cooperation were quite complex. The alternative to cooperation is an adversarial approach to one of the regions most valuable commodities. Cross-border cooperation, in light of perceived security and logistical needs, has been problematic throughout the region. The Chu–Talas project, as it has come to be known, was designed to create a bridge between the two countries, and to facilitate regional cooperation in the management of water resources. The regional situation has been complicated by the peculiar borders that came into existence when the two Republics became independent. This was simultaneously politically difficult and technologically challenging, given the complexities of the issues and the sense of entitlement of citizens in the respective countries. For this reason, the OSCE began a cooperative venture, along with other interested international organizations and donors, to provide an extended program of legislative and technical support.

3. Composition of the Project

The involvement of OSCE has been present throughout the project, and was first presented publicly in April, 2003, at the initial meeting of the project participants in Bishkek, Kyrgyzstan. This was followed by a series of broad workshops and seminars including representatives of technical and political authorities from Kazakhstan and Kyrgyzstan, as well as participating international organizations and donors. Through these workshops the legislative framework for cooperation was developed. Importantly, this process was based upon cooperation and consensus, without which the eventual ratification of the relevant legislation would have been impossible.

The project consisted of four main components:
- The first was to develop the statutes and regulations for the Bilateral Commission on the Chu and Talas rivers.
- The second component called for allocation of costs for operation, management, and rehabilitation of selected water control projects, or the outlining of the financial viability of the Commission.
- The third component of the project involved public participation.
- The final component consisted of a public awareness campaign.

4. Public Participation and Awareness Raising

I would like to dwell a bit on these important components of the project since its enforcement helps to secure democratic trends in the Central Asian region and to foster economic development.

The subprojects on involvement of civil society into decision making process in transboundary water management are as follows:

- Identification of the major stakeholders in the transboundary water region on the issues of transboundary water management of Talas and Chu rivers and assistance to the stakeholders in the communication and in raising their capacity.
- Promotion of the awareness of the Commission and water authorities in Kyrgyzstan and Kazakhstan about interests and needs of the local stakeholders concerning water management in the region.
- Organisation of regular consultations with the relevant authorities for establishing a procedure for the involvement of stakeholders in the process of preparation and implementation of water management decisions.
- Development of the proposals for public participation in management of the transboundary waters in the longer-term future, targeting at creation of the Basin Council and development of a public participation plan as a part of plans under the Commission.
- Dissemination of the information on the project and the Commission activities in the region and internationally.

The public participation component, which is completed, involved significant assistance from new partners, namely the Peipsi Center for Transboundary Cooperation from Estonia, the Zhalgas – Counterpart and the Counterpart – Sheriktech from Kazakhstan and Kyrgyzstan, respectively. The Government of Estonia was very helpful during this phase, supporting the previous experiences of the Peipsi Center in facilitating public information and participation in the Russo–Estonian Lake Peipsi/Chudskoe water basin made the organization a sensible partner in implementing the public participation phase of the Chu–Talas project.

A stakeholder needs assessment was completed, with the intention of identifying the topics and aspects of most interest to the various groups of stakeholders. This assessment included a series of field-based focus groups, where experts were able to gain valuable insights from the individuals most affected by the project (the copies of report you can get at the OSCE Centre). Throughout this process the international organizations involved maintained a strong interest and involvement in the project.

Round table discussions were later held in various locations and included the presence of national and international experts on water management, and the Chu–Talas water management project in particular. Although all efforts have been made to announce developments in the project throughout, a complimentary public awareness campaign was also created to ensure transparency and the availability of information. To this end, a comprehensive website was developed, www.talaschu.org. The website has made available information about the entire process of formulating the Bilateral Commission, and many official documents are freely available. Moreover, relevant informational brochures and materials have been published in Russian, Kazakh, Kyrgyz, and English for public distribution through local and national authorities and NGOs.

5. Chronology of Chu–Talas Events and Initiatives

Date	Activity
January 2000	The Agreement between Kazakhstan and Kyrgyzstan is signed on the water utilities management at the Chu and Talas rivers
February 2000	The Agreement came into force
Beginning 2003	The project on "Creation of a Commission between Kazakhstan and Kyrgyzstan on the Chu and Talas rivers" was launched
November 2003	Draft Statute of the Bilateral Commission is sent for consideration to the Governments of Kazakhstan and Kyrgyzstan
Beginning– End 2004	ADB implementation "Development of efficient complex management of the water resources in two pilot projects in sub basins of the Aral Sea. Pilot project for Chu–Talas"
May 2004	OSCE/UNECE/UNESCAP project launching on public participation (Counterpart Consortium in Kazakhstan and Kyrgyzstan, Centre on Transboundary Cooperation "PEIPSI", Estonia
18–24 June 2004	Under the frame of project implementation on public participation a Study tour is conducted for the representatives of Commission and civil society to Estonia to familiarize with the best practices of the Russian–Estonian Commission on water management

20 August 2004	The composition of Commission is discussed and approved by Kazakhstan and Kyrgyzstan
3 November 2004	The Statute is approved by the Kyrgyz Republic
May 2005	The project on public participation is completed
5 May 2005	The Kazakh Government submit to the Kyrgyz Government for consideration revised draft of the Commission Statute
20 May 2005	The Kyrgyz Government officially requests the Kazakh authorities to adopt the Statue of Commission rather than Sub Commission as the Kazakh side proposed earlier
June–December 2005	Finalization of the Commission's Statute by both countries
December 2005	The Statute of Commission is approved at the Taraz Round Table
Second half of July 2005	Inauguration of Commission. Regional Conference for Central Asian countries for experience replication

6. Policy Challenges

Understanding the complex sensitivities involved, the OSCE and the other partners on the Chu–Talas took an inclusive approach to addressing political and policy matters. A local expert from each country was hired to assist in project implementation, and to ensure that the needs of both countries were adequately represented in each aspects of the project. The local experts were familiar with the technical and political situation in their respective countries, and worked closely with OSCE staff, the staff of partner organizations and international experts to formulate an agreement on the division of resources. This included consultation with each country on the structure of the international agreement, as well as the structure of domestic legislation that would be required for successful implementation of that agreement. It is evident that the additional efforts associated with negotiation were both necessary and effective, as the Commission enjoys strong political support and has very bright prospects, steeped in policy, legislation, and opinion.

Technically, the project was less complicated. The infrastructure existed for water management in Kyrgyzstan. The problem was simply that sizable financial resources would be required to maintain existing water management infrastructure in Kyrgyzstan. It would clearly be inappropriate for Kyrgyzstan to be expected to maintain water management facilities for the use of separate sovereign states. It was, therefore, simply a matter of understanding the needs of both countries, and what those countries could offer to the Bilateral Commission.

7. Project Successes and Possibilities to Exist

With support from the major donors, Great Britain and Sweden, and with additional support from Estonia, OSCE was able to realize significant successes in the Chu–Talas project, even though no comparable cooperative project exists in Central Asia. On a basic level, this project and the Bilateral Commission of which it supported the creation have achieved the long-term likelihood of stable, mutually-beneficial and neighborly exchanges of resources between Kazakhstan and Kyrgyzstan, and that is no small accomplishment. By ensuring stability in this area, it allows local authorities and bureaucrats to focus less upon creating contentious annual agreements, and more upon fostering prudent mechanisms for regional governance.

This project has demonstrated that cooperation can function, and that cooperative agreements of the sort pioneered in this project can and do have significant benefits for all participants. The Chu–Talas project's successes are many, and lie primarily in the steps taken to coordinate water resources in a methodological and stable manner between Kazakhstan and Kyrgyzstan. It would be shortsighted, however, to allow these immediate gains to be the sole benefits gained from the Chu–Talas project. In fact, there are lessons to be learned well beyond the borders of Kazakhstan and Kyrgyzstan, and even beyond Central Asia, on the promises held in well-executed transboundary projects. As this project was aided by the example of the Russian and Estonian partnership in the Lake Peipsi/Chudskoe region, so too might the example created by this project be utilized elsewhere.

The OSCE can work together with politicians and bureaucrats in our member states to draw attention to issues requiring transboundary cooperation, such as water management, and can work together with authorities throughout the successive processes of building transboundary agreements and mechanisms. This is a unique niche of the OSCE, and the "Creation of a commission between Kazakhstan and

Kyrgyzstan on the Chu and Talas rivers" project is evidence of just how successful such efforts can be.

The implementation of this project is coming to its end and it not only contributes to an improved cooperation between Kazakhstan and Kyrgyzstan on the Chu and Talas rivers, but is also be an example for Central Asia on improved cooperation on transboundary waters.

MECHANISMS FOR IMPROVEMENT OF TRANSBOUNDARY WATER RESOURCES MANAGEMENT IN CENTRAL ASIA

DUSHEN M. MAMATKANOV
*Institute of Water Problems and Hydropower of the
National Academy of Sciences
Kyrgyz Republic*

Abstract: The methods of water-management inherited from the Soviet era continue to operate in the Central Asian region. The application of this water-management system is becoming impossible due to intensive development of processes of statehood and market transformations. The approach of compensation using negative profits and losses can be realized through the introduction of the economic mechanisms of transboundary water resources management. This approach was developed by The Institute of Water Problems and Hydropower of the National Academy of Sciences of the Kyrgyz Republic.

Keywords: water resources management, transboundary water resources, economic mechanisms, Central Asia, Kyrgyzstan

1. Introduction

A unique water-energy system was established in Central Asia during the Soviet period. It was generally aimed at maximizing economic benefits from land, water and energy resources of the region. Such approach to natural resources resulted in flooding lands and construction of reservoirs in mountain areas. To recoup lost profits borne by mountain republics, namely by Kyrgyzstan and Tajikistan, due to these works regular food deliveries were arranged. In addition, the induced decrease in electric power generation during winter months was compensated by stable supplies of mineral energy resources to these two republics. Since the breakup of the Soviet Union in the early 90s, the created system has been used by five independent states. Nowadays, it continues to work, however each state of the region aspires to receive the highest possible benefits from the use of natural resources and the system. As a result, upstream states of a mountain zone bear losses associated with river runoff regulation and water supply to neighbors. At the same time, downstream states receive essential benefits from these water supplies and do not participate in operation/maintenance cost sharing. Such situation became possible because political transformations in the region, taken place in the 90s, did not lead to any significant changes in water- management: the Central Asian water allocation scheme developed in the Soviet period has been and is still in practice.

The monopolization of natural resources and introduction of world market prices for some energy resources drove upper riparian states, which lack gas, oil, and coal, into a difficult situation. Attempts to address such a crisis situation by increasing hydroelectric production during the winter time have led to political tensions with downstream neighbors that suffered from floods caused by increased water releases. Unfortunately, the established Interstate Coordination Water Commission (ICWC) proved to be unable to develop effective solutions that would meet new political and economic realities of Central Asia.

There was an urgent need for a strategy of joint actions to address the uneasy situation. Unlike ICWC, the newly created Interstate Council on the Aral Sea Basin was quite effective at that period of time. It facilitated the development of several important agreements and programs, which if implemented, could solve water issues of the region. The following initiatives were of particular importance:

1. The Agreement on Ecological Improvement and Maintenance of Social and Economic Development of the Aral Region was signed by the heads of the states of Central Asia on 26 March 1993 in Kyzyl–Orda. In Article 1 of this Agreement, the signatories recognize the need for the development of a regulating system and improvement of water use in the basin; the development of corresponding interstate legal and the statutory acts; and the introduction of regional general principles of compensation resulting from negative profits and losses.

2. The Program of Concrete Actions on Improvement of the Ecological Situation in the Aral Sea Basin for the Nearest 3–5 Years in View of Social and Economic Development of the Region. This program was adopted by the heads of the states of Central Asia and the Government of the Russian Federation on 11January 1994 in Nukus.
3. The Statement of the Republic of Kazakhstan, the Kyrgyz Republic and Republic of Uzbekistan on Water-Energy Resources Use was signed of 6 May 1996 in Bishkek. This statement underlines "...the necessity to speed up the development of a new strategy of water allocation and economic mechanisms in the field of water-management."

From the analysis of these initiatives, it becomes evident that there would be no serious disagreements between upstream and downstream co-riparian states, if two important issues would have been settled. These issues are:

- Development of a new strategy of the regional water allocation.
- Compensation mechanisms for negative profits and losses.

2. Economic Value of Water Resources

The methods of water resources management inherited from the Soviet era continue to exist in Central Asia and are fixed in the 1992 Almaty Agreement signed by the Republic of Kazakhstan, Kyrgyz Republic, Republic of Uzbekistan, Republic of Tajikistan and Turkmenistan. The existence of these methods contradicts to the present political realities as well as to ongoing market transformations in the region. For instance, the Kyrgyz Republic, within the limits of the Aral Sea basin, is a source of about 30 km^3 of water per year. Despite this fact, Kyrgyzstan is allowed to use only 4 km^3 for irrigation of 465,000 ha, while the country has a potential to increase its irrigated lands to 1,300,000 ha. The current area of irrigated lands in the Kyrgyz Republic is not adequate to provide its food security for the country's population as well as for its future generations. Therefore, a new strategy of water- management and allocation is needed urgently.

The problem of compensation of negative profits and losses can be realized through introduction of the economic mechanisms of transboundary water-resources management. This approach was developed by The Institute of Water Problems and Hydropower of the National Academy of Sciences of the Kyrgyz Republic.

Water resources are natural resources. They; like coal, oil, gas, land, forest funds; represent the elements of nature and are vital resources to support human beings.

Natural resources are included among the public wealth of any state within its borders, and they can become a form of commodities under the conditions of capitalization in market economy. In economic terms goods are defined as:

- Products of work made for exchange and sale.
- Products of productive and economic activity in a physical form.
- Physical objects or services that can be sold on market.

Goods possess two basic features:

- Consumer value (value in use) which determines a measure of need for the goods, its utility and necessity.
- The cost that represents the labor embodied in such goods.

Following these definitions, water resources can be considered as goods particularly for the following reasons:

1. After the disintegration of the Soviet Union into a number of sovereign states, each state declared its right to possess and control public wealth, including natural resources.
2. Each state builds up its relations with neighboring states on principles of a market economy, when natural resources have acquired a form of commodity.
3. Water resources as goods are products of productive and economic activity of upstream states that spend funds from their budget to take measures associated with water formation (reproduction, monitoring, and observation) river runoff regulation, operation of water objects of interstate importance, and water supply to all water-consumers (states).
4. Two basic properties of the goods are inherent for water resources:
    - Consumer cost (necessity and utility) which is the highest in comparison with other natural resources because any form of life is impossible without water.

- Cost or expenses of activities to secure reproduction of water resources, their delivery to water-consumers, and regulation of river runoff.

The above mentioned demonstrates that there are enough reasons to consider water resources as goods that have their economic value.

3. Three Levels of Economic Estimation of Water Resources

It is necessary to emphasize that water resources vary in time and are dynamic in nature. This requires a special approach in defining economic parameters of water as natural resource and as a raw material used in industry and agriculture (note: irrigation is the main consumer of water resources in the countries of Central Asia). Water, together with the economic and social importance, has a high ecological value. Indeed, water, as a source of hydropower, has an advantage being the cheapest, non-polluting and renewable. In addition, management of water can prevent certain negative externalities, namely flooding phenomena.

The developed economic mechanism of transboundary water resources management includes the methodology for various tariffs for water and losses caused by the construction of the Toktogul Hydropower Plant (HPP) and its operation in the irrigation mode [1].

There are three levels of economic estimations for water resources. These levels have to do with:

- River runoff formation.
- Water distribution and supply with the use of the main and inter-economic channels.
- Water supply with the use of water regulation constructions – water objects used for irrigation and power-irrigation purposes.

The first level is related to surface water resources in river runoff formation zones, from river sources up to heads of intake facilities and constructions. At this stage, when water is presented in its natural condition, it has already its economic value as a state provides certain organizations with funds to conduct monitoring of water resources, management, scientific research, anti-flood and anti-mudflow measures, river banks protection works and reproduction of wood plantings. These types of actions and works have been taken into account in "Methods to define tariffs for water as a natural resource".

The average tariff for water as a natural resource (NR) within the republic (T_{rep}^{NR}, som/thousand m³), is calculated with the use of the formula:

$$T_{rep}^{NR} = \frac{\sum_{i=1}^{n} Z_i}{\sum W_{wf}}, \text{ som/thousand m}^3, \quad (1)$$

where:

n – number of budget organizations engaged in water resources (WR) management;

$\sum_{i=1}^{n} Z_i$ – annual expenses (in som) of budget organizations engaged in water resources (WR) management;

$\sum W_{wf}$ – total volume of water formed (wf) in the republic per year, thousand m³.

The annual expenses of i budget organization (Z_i) engaged in WR management, are:

$$Z_i = [Z_{yr,i} + K_{reg,i}(\alpha_{cur} + \alpha_{rep} + \alpha_{ren})] \cdot k_{wr} \cdot P_i, \text{ som} \quad (2)$$

where:

$Z_{yr,i}$ – annual operation expenses of i budget organization under the estimate for a given year (without taking into account expenses for the current, the capital repairs and deductions on renovation), som;

$K_{reg,i}$ – regenerative cost of the basic production assets of i budget organization engaged in WR, for a given year, som;

$\alpha_{cur}, \alpha_{rep}, \alpha_{ren}$ – norms of deductions on the current, capital repairs, and renovation, respectively;

k_{wr} – factor to expenses for formation of WR;

P_i – specification of the conditional profit of i budgetary organization.

This tariff is established for all water-consumers and water-users irrespective of departmental subordination and ownership type. The payments should be transferred to a special state account and should be used purposefully for financing expenditures of the corresponding state budget organizations.

The second level is related to water intake, distribution and water supply with the use of the main and inter-economic channels. All works on water intake and water delivery to the points of water outlet are carried out by the operating water-economic associations (WEA). In Kyrgyzstan, they conduct regional water economic management. A specific tariff approach for WEA services has been developed in "Methods to define tariffs for WEA services related to water supplies to customers". This tariff system has been applied to interstate water-consumers.

This tariff is established for intra-state water consumers, and a methodology of its calculation is not given here.

The third level is related to water supply with the use of water regulation facilities and constructions.

In the Kyrgyz Republic, water basins of long-term and seasonal runoff regulation have been built and work. Some of them (Toktogul, Orto-Tokoi, Kirov, Papan) are the objects of interstate importance which serve water-users of Kyrgyzstan, Kazakhstan, and Uzbekistan. To estimate water-related expenditures, "Methods to define tariffs for water supplied by irrigation and complex water basins" have been developed.

The interstate tariff for water is established for the states, adjacent to Kyrgyzstan, which use the country's water resources; the tariff is calculated as a sum of tariffs for water as a natural resource and tariffs for water regulated by irrigation and complex water objects.

The tariff for services on water supplies from one-target irrigation water basin to neighboring states ($T_{n,irr}^{serv}$, som/thousand m³) can be calculated using the formula:

$$T_{n,irr}^{serv} = \frac{[O_{oper} + C_{ins} + K_n^{irr}(\alpha_{cur} + \alpha_{rep} + \alpha_{ren})] \cdot P_n}{W_{n,irr}^{VP}}, \qquad (3)$$

where:

$T_{n,irr}^{serv}$ – tariff for services associated with water supplies from the nth one-target irrigational water basin to neighboring states, som/thousand m³;

$O_{oper.}$ – annual operation expenses, including salary payments, administrative fees, etc., except the expenses for the current, capital repairs and renovation, som;

C_{ins} – annual expenses for creation of insurance money resources for the shallow periods and material resources for liquidation of emergencies, som;

K_n^{irr} – regenerative cost of the nth irrigational water basin;

α_{cur}, α_{rep}, α_{ren} – accordingly, the norm of deductions for the current, capital repairs and renovation in shares of unit;

P_n – specification of the profit;

$W_{n,irr}^{vp}$ – annual volume of water supplies from the nth one-target irrigational water basin minus sanitary and ecological flushes, thousand m³.

The tariff for services on the water delivery adjustable in complex hydro units (CHU), located on transboundary rivers, ($T_{n,CHU}^{Serv}$, som/thousand m³) is calculated with the use of the formula:

$$T_{n,CHU}^{Serv} = \frac{[O_{oper} + C_{ins} + K_n^{chu}(\alpha_{cur} + \alpha_{rep} + \alpha_{ren})]\gamma_{irr} \cdot P_n}{W_{n,CHU}^{serv}}, \quad (4)$$

where:

$O_{oper.}$ – annual operation expenses (in som), including salary payments, administrative fees, etc., except for expenses for the current, capital repairs and renovation (insom;

C_{ins} – annual charges for creation of insurance money resources for the shallow periods and material resources for liquidation of emergencies, som;

K_n^{CHU} – regenerative cost of any complex hydro unit, som;

α_{cur}, α_{rep}, α_{ren} – accordingly, the norm of deductions for the current, capital repairs and renovation in shares of unit;

γ_{irr} – share of annual operation expenses of N complex hydro unit, carrying on irrigation;

P_n – specification of the profit;

$W_{n,CHU}^{serv}$ – annual volume of water supplies from the nth complex hydro unit minus sanitary and ecological flushes, thousand m³.

The interstate (IS) tariff for water (T^{IS}, som for thousand m³) is established for the states, adjacent to Kyrgyzstan, which use the country's own water resources, and is calculated as the sum of the tariff for water as a natural resource and the tariff for the water adjustable by irrigational and complex water basins.

$$T^{IS} = (T^{NR}_{rep} + T^{SERV}_n)k_{VAT}, \qquad (5)$$

where:

T^{NR}_{rep} – average tariff for water as a natural resource within the republic, som/thousand m³ is calculated with the use of the formula (1);

T_n^{serv} – tariff for services associated with water supplies from the nth complex or one-target irrigational water basin, som/thousand m³, is calculated with the use of the formulas (3,4);

K_{VAT} – factor taking into account the value-added tax.

The developed method is based on the "expense-normative principle" according to which the expenses are formed in terms of basic pricing elements, including operation expenses and normative profits of budgetary organizations providing water management. The expenses related to amortization and operating repairs are taken into account not on the fact, but are calculated in accordance with norms; this is a distinctive attribute of the expense-normative principle. Such approach enables water-economic organizations to receive sufficient funds in the form of a fixed capital to carry out current and capital repair works.

This approach was approved by the representatives of all Central Asian states during the implementation of the USAID project on pricing in water use of Central Asia (1995–1997).

Besides, the tariffs defining an individual share of each state (as a water consumer) for given water object have been developed.

The principle of compensation of the cost of negative consequences is incorporated in the procedure of calculation of annual damage from the creation and operation of the Toktogul HPP. The negative consequences are connected with: flooding; under-flooding of the lands; underproduction of the electric power during the winter time because of the depletion water basins under the irrigational mode; ecological damages from the burning of organic fuels at the Bishkek thermal power plant (TPP).

The annual economic damage (D) general size ($D_{general}$, dollars), rendering to Kyrgyzstan from the Toktogul creation and its operation in the irrigational mode is calculated using the formula:

$$D_{general} = D_{flood} + D_{elec} + D_{ecol}, \qquad (6)$$

where:

D_{flood} – annual loss from flooding and under-flooding (fl) associated with losses of agricultural lands and the Toktogul construction

D_{elect} – annual loss from winter electric (el) power underproduction by the Nizhnii–Naryn HPPs cascade due to the Toktogul HPP work in the irrigational mode

D_{ecol} – annual loss due to negative ecological impacts from the Bishkek TPP that burns fossil fuel to cover hydroelectric power underproduction in the Nizhnii–Naryn HPPs cascade during winter.

The annual amount of losses from flooding and under-flooding of the grounds (D_{flood}, dollars) is determined as:

$$D_{flood} = P(S_1 + kS_2), \qquad (7)$$

where:

P – net profit from one structural hectare of the irrigated grounds by manufacture of agricultural crops in examining area

S_1 – area of the flooded agricultural lands, hectare

S_2 – area of under-flooding lands, hectare

k – share of the net profit losses, related to under-flooding lands.

The annual amount of losses from the winter electric power underproduction in the Nizhnii–Naryn HPPs cascade caused by the irrigational operating mode of Toktogul (D_{en}, dollars), is determined as:

$$D_{en} = \Delta C \cdot \Delta E, \qquad (8)$$

where:

ΔC – excess of cost price of one produced K.W.H. by the Bishkek TPP, above the cost price of K.W.H. of Nizhnii–Naryn HPPs cascade, dollar

ΔE – winter underproduction of the electric power by the Nizhnii–Naryn HPPs cascade, compensated by the electric power production at the Bishkek TPP, million kilowatt-hours.

The annual amount of ecological losses caused by the Bishkek TPP during the winter period of electric power underproduction at the Nizhnii-Naryn HPPs cascade (D_{ecol}, dollar) is determined as:

$$D_{ecol} = (d_{atm} + d_{oxigen.} + y_{dr}) \Delta E, \qquad (9)$$

d_{atm} – specific ecological loss from air pollution by harmful emissions from fuel burning

d_{oxigen} – specific loss from consumption of atmospheric oxygen by burnt fuel

d_{dr} – specific loss from an additional death rate (DR) of workers of the Bishkek TPP due to heavy and dangerous conditions.

4. Conclusion

The annual total loss of Kyrgyzstan is estimated as US$154.9 million. The republic is particularly interested in compensation to be paid by Kazakhstan and Uzbekistan for losses resulted from:

- Flooding of agricultural lands to build such large water objects, like the Toktogul Reservoir.
- Carrying out maintenance works at water objects.
- Keeping Toktogul in the irrigation mode during the winter months, when the demand for electric power is particularly high.

Such payments could help the downstream states to avoid the following problems in the future:

- Floods and flooding during the autumn and winter months.
- Shortage of water supplies during the vegetation period.

Kyrgyzstan in turn could use the received funds to purchase mineral energy resources.

The approach mentioned above is effective in comparison with schemes offered by the 1998 Agreement, particularly a barter exchange of the summer electric power for mineral energy resources. Every summer Kyrgyzstan has problems with selling its electricity, which lead to gas, coal, black oil non-deliveries as well as to working cycle failures on the Bishkek TPP.

In conclusion, it is necessary to emphasize that the above mentioned mechanism of transboundary water resources management offered by the Institute of Water Problems and Hydropower is not something new. Rather, it is a well forgotten fundamental essence of the previous decisions, which have been accepted and authorized by the heads of the Central Asian states. We should concentrate our efforts on their implementation.

References

Asanbekov A.T., Mamatkanov D.M., Shavva K.I., Shapar A.K., The economic mechanism of transboundary water resources management and the main items of strategy of interstate water-division,B.Publishing house of the International institute of mountains, 2000, 44 pp.

Water and steady development of the Central Asia: Materials of projects, Regional cooperation on water and energy resources use in Central Asia, (1998) and Environmental problems and steady development of the Central Asia (2000.) B., "Elite", 2001, 178 pp.

SCIENCE FOR PEACE: MONITORING WATER QUALITY AND QUANTITY IN THE KURA–ARAKS BASIN OF THE SOUTH CAUCASUS

MICHAEL E. CAMPANA
Director, Institute for Water and Watersheds
Professor of Geosciences and
Universities Partnership for Transboundary Waters
Oregon State University
210 Strand Agriculture Hall
Corvallis, OR 97331-2208 USA

BERRIN BASAK VENER
Water Resources Program
Utton Transboundary Resources Center and
Universities Partnership for Transboundary Waters
University of New Mexico
Albuquerque, NM 87131 USA

NODAR P. KEKELIDZE
Head, Center of Physical and Chemical Environmental Monitoring
Tbilisi State University 3, Av. Chavchavadze
0128 Tbilisi, Georgia

BAHRUZ SULEYMANOV
Director of Environmental Physics and Chemistry Center- Azecolab,
Institute of Radiation Problems, National Academy of Sciences of Azerbaijan
Baku, Azerbaijan

ARMEN SAGHATELYAN
Director, Center for Ecological-Noosphere Studies of National Academy of Sciences,
Rep. of Armenia, 68 Abovian Str.
Yerevan 375025, Armenia

Abstract: The Kura–Araks River Basin is an international catchment with five countries – Armenia, Azerbaijan, Georgia, Iran, and Turkey – comprising its watershed. About 65% of the basin area (total area = about 188,200 km^2) falls within the South Caucasus countries of Armenia, Azerbaijan, and Georgia. Both rivers head in Turkey, join in Azerbaijan, and discharge to the Caspian Sea. The length of the Kura is about 1,515 km and that of the Araks is approximately 1,070 km. Soviet monitoring projects from the 1950s through the 1980s collected water quality and quantity data, but these projects do not exist anymore and many of the data appear to be unavailable. In addition, after the dissolution of the Soviet Union, not only did information exchange collapse but the Kura–Arkas also became an international river basin with respect to the three South Caucasus countries. Armenia, Azerbaijan, and Georgia jointly utilize the Kura and Araks rivers and share common problems related to water quantity, water quality, and water allocation. But there are currently no treaties among the three riparians governing water quality, quantity, rights. Monitoring and management of transboundary water are complex problems in any region of the world. In the case of the Kura–Araks River System, the situation is complicated by ongoing regional conflict. Further conflict could be exacerbated by water rights, quantity, and quality issues in the basin, so it is imperative that a culture of cooperation and collaboration be fostered. In November 2002 the South Caucasus River Monitoring Project was funded by NATO's (North Atlantic Treaty Organization) Science for Peace Programme and OSCE (Organization for Security and Cooperation in Europe). This project is not a top-down project managed by NATO and OSCE but was conceived, developed, and is managed jointly by individuals from the three countries. Assistance is provided by experts from Belgium, Norway, and the USA. The project's overall objective is to establish the social and technical infrastructure for international, cooperative, transboundary river water quality and quantity monitoring, data sharing, and watershed management among the Republics of Armenia, Azerbaijan, and Georgia. Its specific objectives are to: increase technical capabilities (analytical chemistry and its application to water resources sampling and monitoring, database management, and communications) among the partner countries; cooperatively establish standardized common sampling, analytical, and data management techniques for all partner countries and implement standards for good laboratory practice (GLP), quality assurance (QA) and quality control (QC); establish database management, GIS, and model-sharing systems accessible to all partners via the WWW; establish a social framework (i.e., annual international meetings) for integrated water resources management; and involve stakeholders. Monthly monitoring is conducted for water quantity (discharge) and water quality parameters at 10 locations in each country. Water quality monitoring consists of the usual basic parameters plus heavy metals, radionuclides, and POPs (Persistent Organic Pollutants). These data will be used to construct a simple dynamic simulation model of the watershed, which will form the basis for a more sophisticated management model. The NATO–OSCE project formally ends in October 2007, although all involved are anxious to continue the work beyond that date. The project has been a model of collaboration and cooperation in a region where such traits have at times been in short supply. Not only have valuable data been collected, but collegial professional relationships also have been established among the participants. In the long run, this latter aspect will likely prove to be the most important product.

Keywords: Water monitoring, international basin, water management, social and technical infrastructure

1. Introduction

The South Caucasus region comprises the countries of Georgia, Armenia, and Azerbaijan. The region is bordered by the Black Sea to the west, the Caspian Sea to the east, the Caucasus Mountains and Russia to the north, and Turkey and Iran to the south (Fig. 1). The three countries have a total population of almost 16 million, with Azerbaijan comprising almost 50% of the total (Table 1).

The Kura–Araks (sometimes spelled "Aras") Basin comprises the major river system in the South Caucasus countries of Georgia, Armenia, and Azerbaijan. Both rivers rise in Turkey and flow into the Caspian Sea after joining in Azerbaijan. Of the total basin area of about 188,200 km^2, almost twothirds, or about 122,200 km^2, are in the aforementioned countries; the remaining basin area is in Turkey and Iran.

The water users in all three countries are faced with water quality and quantity problems. In general terms, Georgia has an oversupply of water, Armenia has some shortages based on poor management, and Azerbaijan has a lack of water (TACIS 2003). The main use of Kura–Araks water in Georgia is agriculture, and in Armenia, it is agriculture and industry. In Azerbaijan, the Kura–Araks water is the primary source of fresh water, and is used for drinking water. Almost 80% of the countries' wastewater load is discharged into the surface waters of the Kura–Araks Basins and their tributaries (UNECE 2003). The basin is excessively polluted due to a lack of treatment for urban wastewater and agricultural return flows, pesticides such as DDT that are used in Azerbaijan, and the recent resurgence of chemical and metallurgical industries in Georgia and Armenia (TACIS 2002).

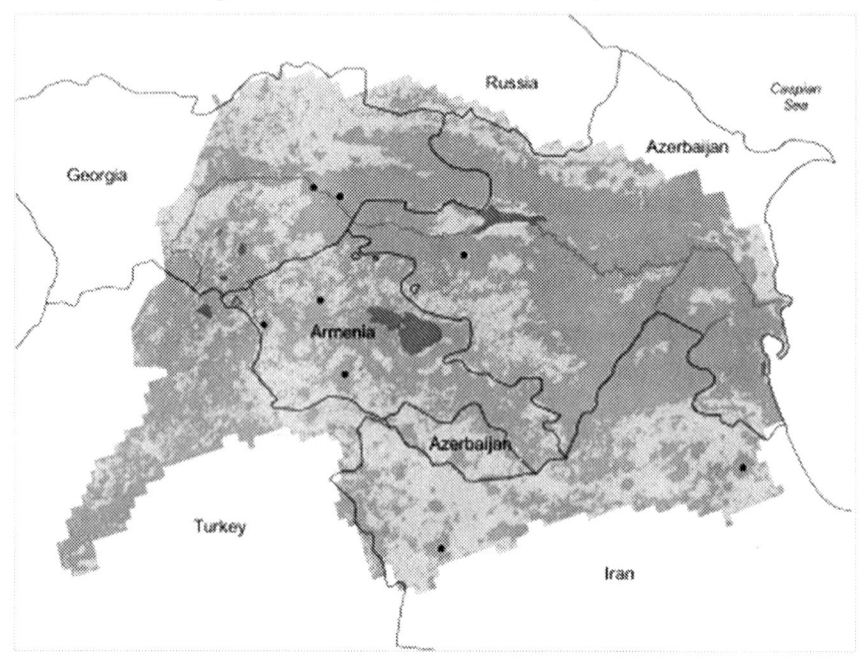

Figure 1. Georgia, Armenia, and Azerbaijan. Kuras–Araks watershed enclosed in blue (Vener, 2006).

2. Water Resources of the Kura–Araks Basin

The Kura–Araks Basin is situated south of the Caucasus Mountains. Its borders are northeastern Turkey, central and eastern Georgia, and northwestern Iran It contains almost all of Azerbaijan and Armenia (Fig. 1).

The Kura River originates in northern Turkey, flows through Georgia and Azerbaijan and then directly discharges into the Caspian Sea. The total length of the Kura River is about 1,515 km and it has an average discharge of 575 million cubic meters per year or MCM/yr (CEO 2002).

The Araks (or Aras) River originates in Turkey and after 300 km forms part of the international borders between Armenia and Turkey, for a very short distance between Azerbaijan and Turkey, between Armenia and Iran, and between Azerbaijan and Iran. The Araks River joins the Kura River in Azerbaijan (TACIS 2003). The Araks River is about 1,072 km long and it has an average discharge of 210 MCM/yr.

Table 1 shows the distribution of watershed area by country; Table 2 shows land use.

TABLE 1. Watershed area of the Kura and Araks Rivers in each country (Vener, 2006).

Country	Population (millions) (July 2003 est.)	Kura River % of total basin area	Kura River Area (km^2)	Araks River % of total basin area	Araks River Area (km^2)
Armenia	3.3	15.79	29,741	22	22,090
Azerbaijan	7.8	30.70	57,800	18	18,000
Georgia	4.9	18.43	34,700	–	–
Turkey & Iran	–	35.06	66,000	60	61,000
Total	16	100.00	188,241	100.00	101,090

(Sources: TACIS 2003, USAID 2002, USCIA 2004).

TABLE 2. Land use in the Kura–Araks Basin (km^2) (Vener, 2006).

State	Land Area	Disputed Area	Forested Area	Arable land JRMP	Arable land CIA	Meadow, pasture	Other
AR	29,800	1,500	4,250	5,600	5,215	8,300	10,091
AZ	86,600	2,000	7,590	15,290	16,714	20,936	12,000
GE	67,700	600	10,900	7,700	7,813	NA	NA

(Sources: TACIS 2003, US CIA 2004)

Table 3 shows that water resources are not distributed equally in the South Caucasus. While Georgia has more water than it needs, Azerbaijan is left with a water deficit; furthermore its groundwater is of poor quality. It obtains 70% of its drinking water from the Kura–Araks rivers. Armenia has a surface water shortage but has a large fresh groundwater stock that it uses for drinking water (TACIS 2003).

TABLE 3. Kura–Araks Basin average annual water balance (km^3) (Vener, 2006).

	AR	AZ	GE
Precipitation	18	31	26
Evaporation	(11)	(29)	(13)
River Inflow	1	15	1
River Outflow	(8)	(18)	(12)
Underground inflow	1	3	1
Underground outflow	(1)	(2)	(3)

(Source: TACIS 2003. Parentheses indicate depletion).

Table 3 shows that the most precipitation and evaporation occurs in Azerbaijan followed by Georgia and Armenia in that order.

Water is used for municipal, industrial, agricultural, irrigation, fishery, recreation, and transportation purposes. The main water use is agriculture, followed by industry and households uses (Table 4). Table 3 shows that Azerbaijan has the most arable land followed by Georgia and Armenia and that even though Azerbaijan has the most arable land it is the one that is faces a water deficit (Table 4).

Table 4 indicates that Azerbaijan withdraws 57.9% of its actual renewable water resources, Armenia withdraws 28.2% of its actual renewable water, whereas Georgia withdraws only 5.2% of its actual renewable water. However, as a water resources-rich country Georgia's withdrawal per capita (cubic m) is 635 m^3 while Azerbaijan's is 2,151 m^3, and Armenia's is 784 m^3 (see Table 4). It is evident that per capita water withdrawal is disproportionate to water availability among the three countries.

The main rivers have only two reservoirs but the tributaries have more than 130 major reservoirs. The total capacity of the reservoirs and ponds is almost 13,100 MCM (TACIS 2003).

With respect to storm water and sewage effluent discharges, the Kura–Araks receives 100% of Armenia's, 60% of Georgia's, and 50% of Azerbaijan's.

3. Political, Social, and Economic Landscape

Armenia, Azerbaijan, and Georgia gained their independence from the Union of Soviet Socialist Republics (USSR) in 1991. The South Caucasian states are neither fully democratic nor fully authoritarian states. All three countries attempted to introduce democratic systems, and held relatively free elections in 1990–1992 (SIDA 2002). However, the region reverted to increased authoritarian rule because of the pressures from war, threats of economic collapse, and the countries' inexperience with participatory politics.

A series of ethnic conflicts erupted in Nagorno–Karabakh, Abkhazia, Javakheti, and other regions of the South Caucasus. Because of these internal and international ethnic conflicts the region has about 1,500,000 refugees and/or Internally Displaced Persons (IDP) (SIDA 2002). The South Caucasus region remains in crisis because of ethnic conflicts, poor economies, environmental degradation, and political instability.

Of the three countries, Georgia has made the greatest progress towards building a democratic polity. Azerbaijan and Armenia are still in a transition period from authoritarian regimes to full democracies. Political violence has been a constant threat in the three countries since independence as all have experienced *coup d'états*, insurrections, or attempts to assassinate political leaders (SIDA 2002). As a result, political and socioeconomic reform processes in all three countries have been slow and continually suffer setbacks. Widespread corruption, bureaucratic difficulties, and political instability have continued the South Caucasus' reputation as a relatively high-risk area for business (USDS 2003; SIDA 2002).

TABLE 4. Water resources and freshwater ecosystems in the South Caucasus (Vener, 2006).

Internal Renewable Water Resources (IRWR), 1977–2001	AR	AZ	GE
Surface water produced internally	6.2	6	57
Groundwater recharge (cubic km)	4.2	7	17
Overlap (shared by groundwater and surface water) (cubic km)	(1.4)	(4)	(16)
Total internal renewable water resources (surface water + groundwater – overlap) (cubic km)	9	8	58
Per capita IRWR, 2001 (cubic meters per person)	2,393	995	11,151

(Continued)

Internal Renewable Water Resources (IRWR), 1977–2001	AR	AZ	GE
Natural Renewable Water Resources (includes flows from other countries)			
Total, 1977–2001 (cubic km)	11	30	63
Per capita, 2002 (cubic meters per person)	2,778	3,716	12,149
Annual river flows:			
From other countries (cubic km)	1	21	8
To other countries (cubic km)	3	-	11.9
Water Withdrawals			
Year of withdrawal data	1994	1995	1990
Total withdrawals (cubic km)	2.9	16.5	3.5
Withdrawals per capita (cubic m)	784	2,151	635
Withdrawals as a percentage of actual renewable water resources	28.2%	57.9%	5.2%
Withdrawals by sector (as a percent of total){a}			
Agriculture	66%	70%	59%
Industry	4%	25%	20%
Domestic	30%	5%	21%
Desalination (various years)			
Desalinated water production (cubic m)	0	0	0
Freshwater Fish Species, 1990s			
Total number of species	41	61	84
Number of threatened species	0	5	3
Freshwater Food Production			
Freshwater fish catch{b}			
1990 (metric tons)	2,698	40,389	117
2000 (metric tons)	1,105	18,795	194
Freshwater aquaculture production			
1987 (metric tons)	-	-	-
1997 (metric tons)	1400	488	1

Notes:

a. Totals may exceed 100% due to groundwater withdrawals, withdrawals from river inflows, and the operation of desalinization plants.

b. Freshwater fish production data refer to freshwater fish caught or cultivated for commercial, industrial, and subsistence use.

c. Parentheses indicate depletion.

(Sources: Modified from EarthTrends 2003a, 2003b, 2003c and FAO/AQUASTAT 2006a, 2006b, 2006c, 2006d)

In a more positive vein, the South Caucasus lies on an ancient trade route known as the "Silk Road". The region acts as a natural bridge between Europe and Asia and is surrounded by three regional powers: Russia, Iran, and Turkey. It has a favorable geographic location at the crossroads of Asia, Europe, and the Middle East. This is one of the reasons the international community began to realize the geopolitical and geoeconomic importance of the South Caucasus in the world (SIDA 2002). Thus, the three states have been eager to develop east–west and north–south transport corridors through their territory, such as the recently completed BTC (Baku–Tbilisi–Ceyhan) oil pipeline, which begins in Baku, Azerbaijan, and terminates at the Turkish Mediterranean port of Ceyhan. In other words, restoration of the ancient Silk Road may help restore the socioeconomic and political stability to the region (SIDA 2002).

Forces leading towards democratization were strengthened by membership in the Council of Europe. The three countries were accepted into the Council of Europe after their independence: Georgia in 1999 and Azerbaijan and Armenia in 2001. Being a member of the Council of Europe is the first step in a country's candidacy for European Union membership, which requires meeting European Union standards and harmonization with the European Union legislation. Ultimately, the European Union membership is the goal of these three countries.

It is important to note that there are constructive involvements in the area from different countries such as the USA and Sweden, and international/intergovernmental organizations (IGOs) such as the European Union (EU), World Bank (WB), North Atlantic Treaty Organization (NATO), United Nations (UN), United Nations Development Program (UNDP), Organization for Security and Cooperation in Europe (OSCE), and the United States Agency for International Development (USAID). These organizations have an investment portfolio and international funding and credits available specifically for projects related to the environment, energy, communication, and education.

4. Hydropolitics

4.1. INTRODUCTION

During the Soviet era, each country was within the USSR sphere and water resources management of the basin was contingent upon the policy that the USSR was implementing at the time. When they became independent states, the three countries had neither water resources management regulations nor water codes. However, each country has adopted water codes within the last 15 years: Armenia in 1992 and revised in 2002 according to the European Union Water Framework Directives (EU–WFD); and Georgia and Azerbaijan in 1997. Nevertheless, there is no uniform control and/or management system for the rivers and, in the post-Soviet period, no water quality monitoring by the riparian countries since 2002.

While the three countries are willing to cooperate on water-related issues, they have not solved their political, economic, and social issues. There are currently no water treaties among the three countries, a condition directly related to the political situation in the region. There is recognition of the importance of water and river basin management, which provides the countries with a good foundation for a transboundary water management agreement.

However, there are political issues which make agreements difficult among the countries. Nagorno–Karabakh is one of the main obstacles, making it difficult for Azerbaijan and Armenia to sign a treaty even though it may relate only to water resources management. Another obstacle is the Javakheti region of Georgia. Ethnic Armenian groups in Javakheti who are seeking greater autonomy and closer ties with Armenia have fomented a confrontation between Armenia and Georgia and a resumption of hostilities in the region (SIDA 2002).

Armenia has not yet signed transboundary water-related conventions, but its Water Code takes into account the transboundary aspects of water. The Water Code also establishes the principles and first steps for river basin management, which is a very important step for transboundary water management.

As a downstream country suffering from a water shortage, **Azerbaijan** is open to signing international water-related conventions (TACIS 2003). Azerbaijan signed and ratified the 1992 Helsinki Convention which governs the protection and use of transboundary watercourses (http://www.unece.org/env/water/), and wants Armenia and Georgia to ratify it. However, in the area of international conventions, Azerbaijan is far behind Georgia and Armenia. Also complicating the situation in Azerbaijan is the ownership of the water which can be the state, municipalities, or the private sector.

Georgia has decided to harmonize its legislation with international development. Georgia has signed more international conventions than Azerbaijan and Armenia and is currently discussing ratifying the 1992 Helsinki Convention. After the Soviet Union's dissolution, the countries in the South Caucasus were primarily faced with environmental degradation stemming from agriculture and industry.

The current water work in the basin will be the first step towards cooperation and it is possible that the three countries will be able to carry this positive spirit into resolving other areas of conflict.

4.2. ONGOING CONFLICTS INFLUENCING TRANSBOUNDARY WATER RESOURCES ISSUES

Although there are a number of ongoing conflicts in the South Caucasus, we will mention only those that are between countries as they would exert the strongest control on any transboundary water issues.

The Nagorno–Karabakh region is predominantly an Armenian-populated area in western Azerbaijan. Armenia supports ethnic Armenian secessionists in Nagorno–Karabakh and militarily occupies Nagorno–Karabakh, 16% of Azerbaijan. After the occupation, more than 800,000 Azerbaijanis were forced to leave the occupied lands; another estimated 230,000 ethnic Armenians were forced to leave their homes in Azerbaijan and flee into Armenia (USDS 2003, CIA 2004). A cease-fire between Armenia and Azerbaijan was signed in May 1994 and has held without major violations ever since. The "Minsk Group," part of OSCE, continues to mediate disputes.

Javakheti is an area that is part of Georgia bordering Turkey, and has a total population of 100,000 people. Almost 90% of the population is Armenian. Thus, Javakheti is often cited as a secessionist region (NIC 2000). The region is more integrated with Armenia than Georgia. Armenia supports demand for local autonomy of the region.

5. Water Projects in the South Caucasus

The Kura–Araks Basin is the one of the most highly-desired areas for foreign organizations, and there are several projects related to the management of the basin's water resources. Major regional projects related to transboundary water resource management are: the EU TACIS Joint River Management Project (TACIS JRMP) in cooperation with UNDP, the NATO–OSCE South Caucasus River Monitoring Project and USAID's South Caucasus Water Management Project (see Vener, 2006, Appendix II, for a comprehensive listing of all projects as of July 2006). Even though most of the projects are related, there is little or no cooperation among the organizations and agencies. Nearly all the projects have common goals and activities or overlapping actions but they do not share or exchange information due to the lack of legally binding data exchange requirements. The NATO–OSCE project differs from the others in that it is a bottom-up initiative, proposed by individuals in each country, and managed by them with assistance from experts in the USA, Belgium, and Norway.

The sector-based approach to water resources management is still widely used and integrated river basin-based water management principles are not used region-wide. There are some efforts to introduce these approaches and to establish specific water authorities for coordinated water resources management and improved performance in the South Caucasus.

A lack of communication is not only a problem on the national and international levels, but also among the international agencies and organizations. That is why the ongoing projects (OSCE, USAID, TACIS, and NATO–OSCE, etc.) aim at strengthening the cooperation among related agencies at all local, national, regional and inter-organizational levels and demonstrate the effectiveness of integrated water resources management.

Thus, these projects and the lead actors must come together and clearly define their objectives, goals, and activities for more efficient results. It is crucial to establish a coordinating group that includes each project's lead individual and representatives from each country for efficient and more sustainable results. According to the senior author, creation of a regional coordination group was suggested at the "Transboundary Water Issues in South Caucasus" workshop in Tbilisi in November 2002. However, no actions to date have occurred.

6. South Caucasus River Monitoring Project

6.1. CURRENT STATUS

The aforementioned projects and the detailed list presented by Vener (2006) differ from the South Caucasus River Monitoring Project (SCRMP) in that the SCRMP is more a bottom-up project conceived by local people.

Soviet monitoring projects from the 1950s through the 1980s collected water quality and quantity data, but these projects do not exist anymore and many of the data appear to be unavailable. In addition, after the dissolution of the Soviet Union, not only did information exchange collapse but the Kura–Arkas also became an international river basin with respect to the three South Caucasus countries. As van Harten (2002) states, the Kura–Araks is one of the "new" transboundary river systems of the former "Second World" whose problems are largely *terra incognita.*

Armenia, Azerbaijan, and Georgia jointly utilize the Kura and Araks rivers and share common problems related to water quantity, water quality, and water allocation. But there are currently no treaties among the three riparians governing water quality, quantity, or water rights. Monitoring and management of transboundary water is a complex problem in any region of the world. The problem in the case of the Kura–Araks river system is complicated by ongoing regional conflict. Conflict could be exacerbated by water rights, quantity, and quality issues in the basin, so it is imperative that a culture of cooperation and collaboration be fostered.

In November 2002 the South Caucasus River Monitoring Project was initiated, funded primarily by NATO's Science for Peace Programme and secondarily by OSCE. Some funding was also provided by Statoil, the Norwegian state oil company. This project is not a top-down project managed by NATO and OSCE but was conceived, developed, and is managed jointly by individuals from the three countries. Assistance is provided by experts from Belgium (Professor. Freddy Adams, University of Antwerp), Norway (Professor Eiliv Steinnes, Norwegian University of Science and Technology) and the USA (Professor Michael E. Campana, Oregon State University). The project's overall objective is to establish the social and technical infrastructure for international, cooperative, transboundary river water quality and quantity monitoring, data sharing, and watershed management among the Republics of Armenia, Azerbaijan, and Georgia. Its specific objectives are to:

- Increase technical capabilities (analytical chemistry and its application to water resources sampling and monitoring, database management, and communications) among the partner countries.
- Cooperatively establish standardized common sampling, analytical, and data management techniques for all partner countries and implement standards for good laboratory practice (GLP), quality assurance (QA) and quality control (QC).
- Establish database management, GIS, and model-sharing systems accessible to all partners via the WWW.
- Establish a social framework (i.e., annual international meetings) for integrated water resources management.
- Involve stakeholders.

Monthly monitoring is conducted for water quantity (discharge) and water quality parameters at 10 locations in each country. Water quality monitoring consists of the usual basic parameters (major and minor ions, pH, etc.) plus selected heavy metals, radionuclides, and POPs (Persistent Organic Pollutants). Sampling and analytical protocols have been agreed upon by all parties. This will help minimize "complaints" from any of the riparians about someone else's erroneous data. The group holds a meeting at least once per year to discuss current work, equipment needs, planned work, and to present results. The meetings are generally held in Tbilisi, Georgia, since the current political situation between Armenia and Azerbaijan makes Tbilisi the most convenient venue.

Data are posted on the project WWW site www.kura–araks–natosfp.org, which is maintained by Azerbaijan. Posting ensures that all have access to all the data. These data will be used to construct a simple dynamic simulation model of the watershed, which will ultimately form the basis for a more sophisticated river basin management model. The dynamic simulation model itself could be used as the basis for a "shared vision" or "mediated modeling" (van den Belt, 2004) approach to conflict management and the development of a treaty or compact to manage the system.

6.2. FUTURE WORK

The NATO–OSCE project formally ends in October 2007, although all involved are anxious to continue the work beyond that date. It is imperative that the monitoring be continued so that unimpeachable data can be obtained and used as the basis for future agreements. The project has been a model of collaboration and cooperation in a region where such traits have at times been in short supply. Not only have valuable data been collected, but collegial professional relationships also have been established among the participants. In the long run, this latter aspect will likely prove to be the more important product.

The region could benefit from an expanded project that should address the following:

- Groundwater. The current project does not explicitly consider the presence of groundwater, an important source of water and intimately connected with streamflow. A groundwater-surface water simulation model could then be devised.
- Environmental flows and ecosystems needs. Develop requirements for these items.
- Public health monitoring. Ewing (2003) cited the concern over public health related issues (e.g., waterborne diseases, pathogenic organisms), since much untreated sewage is discharged directly into the waters of the Kura–Araks Basin. She also designed a surface water monitoring plan. This plan should be used as a template for public health monitoring.
- Update country water codes. This would consider potential changes in light of new information/changing conditions. The aim should be to manage water quantity, water quality, and ecosystem health simultaneously and transparently.
- Involve Turkey and Iran, the other two riparians.
- Develop an international agreement. This would provide a framework for joint management of the Kura–Araks Basin: surface water, transboundary groundwater, water quality, and ecosystem health.

The project could also serve as a template for other regions, such as Central Asia.

7. Summary

Since the dissolution of the Soviet Union the Kura–Araks River Basin has become a transboundary basin with respect to the former Soviet republics of Armenia, Azerbaijan, and Georgia. No formal water allocation agreement exists, and little water quality monitoring has occurred since the republics became independent and no mechanisms have been devised to manage the waters in the Kura–Araks Basin. To remedy these issues, NATO and OSCE have funded a project enabling all three riparians to monitor surface water quality and quantity on a monthly basis and post data on a transparent WWW site maintained by Azerbaijan. The data will be used to develop a dynamic simulation model of the basin, which could form the basis for a "mediated modeling" approach to the creation of a basin treaty. Although the current project is a much-needed first step, more work must be done, especially with respect to the inclusion of groundwater, ecosystem needs, and public health monitoring.

Political differences among the countries, especially the Nagorno–Karabakh issue between Armenia and Azerbaijan, must be resolved before a meaningful agreement among all three countries regarding water allocation, quality, and ecosystem requirements can be developed and implemented. Any such agreement must also consider Turkey and Iran, the other two riparians in the basin.

But a strong start has been made.

8. Acknowledgements

The authors are grateful to NATO's Science for Peace Programme for supporting project SfP 977991. Special thanks are due to NATO SfP's Dr. Chris De Wispelaere (Director), Dr. Susanne Michaelis (Associate Director), and Ms. Nicolle Schils-van Maris, whose help and understanding have been invaluable. We are grateful to OSCE and Statoil for their support. We also extend our gratitude to those who helped make the SCRMP what it is today: NATO experts Professor Freddy Adams (Belgium) and Professor Eiliv Steinnes (Norway). Mr. Gianluca Rampolla, formerly of OSCE, is also gratefully acknowledged for his support.

References

Caucasus Environmental Outlook (CEO), 2002, Caucasus Environmental Outlook Report, completed through financial assistance provided by UNDP and Swiss Agency for Environment, Forest, and Landscape. Available at http://www.gridtb.org/projects/CEO/full.htm

EarthTrends, 2003a, Water Resources and Freshwater Ecosystems-Armenia. Available at http://earthtrends.wri.org.

EarthTrends, 2003b, Water Resources and Freshwater Ecosystems-Azerbaijan. Available at http://earthtrends.wri.org.

EarthTrends, 2003c, Water Resources and Freshwater Ecosystems-Georgia. Available at http://earthtrends.wri.org.

Ewing, Amy, 2003. Water Quality and Public Health Monitoring of Surface Waters in the Kura–Araks River Basin of Armenia, Azerbaijan and Georgia. Publication No. WRP-8, Water Resources Program, University of New Mexico, Albuquerque, NM. (available at www.unm.edu/~wrp/wrp-8.pdf)

National Intelligence Council (NIC), 2000, Central Asia and South Caucasus: Reorientations, International Transitions, and Strategic Dynamics Conference Report, October 2000. Available at http://www.fas.org/irp/nic/central_asia.html, accessed on 5/7/2004.

Swedish International Development Cooperation Agency (SIDA), 2002, 'The South Caucasus: Regional Overview and Conflict Assessment', Prepared by Cornell Caspian consulting (CCC) under the contract by SIDA. Available at Available at http://www.cornellcaspian.com/sida/sida.html; accessed: 2/26/2004.

Technical Assistance to Commonwealth of Independent States (TACIS), 2003, European Commission Inception Report, Joint River Management Programme (JRMP) of the Kura Basin, Annex 6: Georgia Country Report. Available at http://www.parliament.the-stationery-office.co.uk/pa/ld199798/ldselect/ldeucom/157/15703.htm; accessed: 2/25/2004).

Technical Assistance to Commonwealth of Independent States (TACIS), 2002, Partnership and Trust: The TACIS Program, 157/1570. Available at http://www.parliament.the-stationary-office.co.uk/pa/ld1999798/ldselect/ldeucom/157/1570; accessed: 2/25/2004.

United Nations Economic Committee for Europe (UNECE), 2003, Environmental Performance Review 2003. Available at http://www.unece.org/env/epr/studies/htm; accessed: 7/2/2006.

United Nations (UN) Food and Agriculture Organization (FAO)/AQUASTAT, 2006a, 'Water and Food Security Country Profiles. Water Balance, Water Resources Data, Armenia'. Available at http://www.fao.org/countryProfiles/water/default.asp?search=search&iso3=ARM http://www.fao.org/ag/agl/aglw/aquastat/water_res/armenia/armenia_wr.xls http://www.fao.org/ag/agl/aglw/aquastat/water_res/armenia/armenia.tif

United Nations (UN) Food and Agriculture Organization (FAO)/AQUASTAT, 2006b, 'Water and Food Security Country Profiles. Water Balance, Water Resources Data, Azerbaijan'. Available at http://www.fao.org/countryProfiles/water/default.asp?search=search&iso3=AZE http://www.fao.org/ag/agl/aglw/aquastat/water_res/

azerbaijan/azerbaijan_wr.xls http://www.fao.org/ag/agl/aglw/aquastat/water_res/azerbaijan/azerbaijan.tif

United Nations (UN) Food and Agriculture Organization (FAO)/AQUASTAT, 2006c, 'Water and Food Security Country Profiles. Water Balance, Water Resources Data, Georgia. Available at http://www.fao.org/countryProfiles/water/default.asp?search=search&iso3=GEOhttp://www.fao.org/ag/agl/aglw/aquastat/water_res/georgia/georgia_wr.xls http://www.fao.org/ag/agl/aglw/aquastat/water_res/georgia/georgia.tif

United Nations (UN) Food and Agriculture Organization (FAO)/AQUASTAT, 2006d, 'Water and Food Security Country Profiles. Glossary of terminology used in the water resources survey. Available at http://www.fao.org/ag/agl/aglw/aquastat/water_res/indexglos.htm

US Agency for International Development (USAID), 2002, Mission for the South Caucasus, Water Management in the South Caucasus Analytical Report: Water Quantity and Quality in Armenia, Azerbaijan and Georgia, prepared by Development Alternatives, Inc. for USAID.

US Central Intelligence Agency (CIA), 2005, Factbook, Country Profiles: Azerbaijan, Armenia and Georgia. Refugees and Internally Displaced Persons (IDPs) Available at https://www.cia.gov/cia/publications/factbook/docs/profileguide.html; accessed: 04/07/2004.

US Central Intelligence Agency (CIA), 2004, Factbook, Country Profiles: Azerbaijan, Armenia and Georgia. Available at https://www.cia.gov/cia/publications/factbook/docs/ profileguide.html; accessed: 04/07/2004.

US Department of State (USDS), 2003, U.S. Government Assistance to and Cooperative Activities with Eurasia, Bureau of European and Eurasian Affairs. Available at http://www.state.gov/p/eur/rls/rpt/23603.htm; accessed: 4/13/2004.

van den Belt, Marjan, 2004. Mediated Modeling: A Systems Dynamics Approach to Environmental Consensus Building. Washington, DC, Island Press, 296p.

van Harten, Marten, 2002. Europe's troubled waters. A role for the OSCE: the case of the Kura–Araks. Helsinki Monitor 13(4):338–349.

Vener, Berrin Basak, 2006. The Kura–Araks Basin: Obstacles and Common Objectives for an Integrated Water Resources Management Model among Armenia, Azerbaijan, and Georgia. Master of Water Resources Professional Project, Water Resources Program, University of New Mexico, Albuquerque, NM.

HAZARDOUS POLLUTANT DATABASE FOR KURA–ARAKS WATER QUALITY MANAGEMENT

BAHRUZ SULEYMANOV, MAJID AHMEDOV,
FAMIL HUMBATOV, AND NAVAI IBADOV
Institute of Radiation Problems of National Academy of Sciences
Baku, Azerbaijan

Abstract: Historically, there has been a deficiency of data for important chemicals, such as heavy metals, radionuclides and pesticides contents from Kura–Araks river waters. A NATO–OSCE 977991 project began to fill such information gaps as well as develop a scientifically based platform for transboundary water quality management issues. The results of 2 years monthly studies of 12 heavy metals, Ag, As, Cd, Cr, Co, Cu, Hg, Mn, Mo, Ni, Pb, Zn are presented. These data for determination of the ranges of radionuclides and pesticides and PCBs have also been initiated of new approach for processes which control of water quality in Azerbaijan parts of the Kura and Araks rivers.

Keywords: water quality management, heavy metals, radionuclides, pesticides, PCBs, PAH, Kura–Araks, Azerbaijan

1. Introduction

The South Caucasus region is one of the most unique places for environmental chemist and geochemist in the world from the perspective of geography, geology, and natural/artificial affects [1]. Situated between Russia to the North and Iran and Turkey to the South, this region is under the influence of the Black Sea and Caspian Sea hydrometeorology processes.

Although the Kura–Araks basin is listed as number 53 in terms of the largest world's river basins, only 13 river basins cover areas of 5 or more countries. And only a few river basins cover almost all 3 countries in their flow. It is well known that 80% of Azerbaijan territory lies within of is affected by the Kura–Araks watershed. In the meantime Kura and Araks rivers and its tributaries are natural drainage system for 100% of Armenian storm and sewage waters and 52% of Georgian waters. Mean 30×10^9 m^3 waters which formed in Turkey, Georgia, Armenia, Iran and Azerbaijan goes to Caspian Sea annually. Most of these waters are considered as surface water and are using intensively both for agricultural, municipal and industrial usage. Azerbaijan uses Kura water also for drinking which makes river water quality items very sensitive to the needs of the community. Overall 900 km Kura flows from Georgia boundary till Caspian Sea through Azerbaijan needs a full scale study to understand the fates of hazardous pollutants, to control of water quality and adequate planning of water usage.

Figure 1. Kura–Araks watershed

Intensive development during 1950–1990 of the mining industry (Armenia), metallurgic (Georgia), chemical, power and processing industries (in all of 3 republics), and irrigated agriculture resulted in dramatically increasing of both water catchment area, sewage into rivers and finally with significant deterioration of water quality. According to official data about 300 million cubic meters of polluted sewage was discharged annually into the river basin within the Armenian territory and 265 million cubic meters in Georgia. Although there are not large industrial center on Kura in Azerbaijan territory, rivers and tributaries flow thorough the populated agriculture areas and undergo stresses from related activities.

It is expected of decreasing concentrations of series of pollutants in Kura and Araks waters because of industrial collapse of Armenia and Georgia. Also at this time, farmers began to use fewer amounts of chemicals which also led to decreasing amounts of total pollution in river waters.

Water quality management requires a comprehensive database for each of targeted hazardous pollutants. One of the original scientific programs on worldwide river quality was initiated in the mid-1950s by the Association for Scientific Hydrology [2]. Livingston [3] effecttively mined these data compilations and created the first truly globalscale treatise on river chemistry.

Routine water quality monitoring had started during 1950s and the advances in spectrometry and chromatography allowed for research programs in western countries beginning from 1980. The worldwide freshwater quality monitoring network – GEMS-WATER – launched in the late 1970s by UNEP, WHO, UNESCO, and WMO established a global-scale database on the quality of rivers. Lists of collected data cover fates of series of pollutants between 20 till 100s for some river basins. In the meantime in territory of Former Soviet Union (FSU), where equipments based on old technology were used, we had a deficiency of reliable database since the start of the 21st century. These data do not allow for the use of computer modeling software developed between 1990 and 2000 to derive of fates of main pollutants in FSU's rivers.

Each water modeling is based on standard aqueous chemistry mechanisms. Scientists need to determine the true concentrations of chemical traces by applying of adsorption, sedimentation, leaching and other factors. Again this requires the determination of concentrations of chemicals in water samples. Basic level of chemical in groundwater compartment can be determined by geologic patterns of the aquifer. Storm waters from rain and snows could broaden the ranges of chemicals in natural waters, and in urban areas artificial affect may compete with natural inputs.

In 2000, when western specialists had collected previous existent data and tried to apply modeling for Kura–Araks waters it became obvious the deficiency of reliable water content data.

2. Project Objectives

Tracing for 4 types of pollutants [4–7] are urgent for Azerbaijan stakeholders:

1st type relate of organic compounds:
- Pesticides, which washing and leaching from agricultural areas.
- PCBs- These compounds had used in energy system.
- PAHs- These comes from crude oil products.

These compounds are known as persistent organic compounds (POPs) which monitoring is important part of world environmental program.

2nd type compounds are heavy metals. Such hazardous inorganic as As, Cd, Cu, Cr, Hg, Ni, Pb, Mo could be produced from mining and metallurgy plants.

3rd type is radionuclides, which are of interest for stakeholders.

- Natural radionuclides are formed as a result of leaching in geology rocks and mainly consist of ^{238}U and the daughter products of ^{232}Th (^{228}Ra, ^{226}Ra, ^{224}Ra and Rn isotopes). Ratio of isotopes could be used also for calculating of surface and ground water mixing percents.

- Artificial radionuclides. Again additional Radium isotopes could be leaching during mining processes. Additionally Cs isotopes precipitate from the atmosphere as residues of nuclear bomb testing and Chernobyl catastrophe.

4th type is related to microbiology contents, especially in the context of usage of Kura water for drinking in Azerbaijan.

In 2002 with support of NATO and OSCE, the Science for Piece Project SfP977991 had started which was intended to develop a methodology for gathering reliable water quality data for Kura–Araks area. Modern equipment, trainings were supplied for the establishment of 3 labs in Azerbaijan, Armenia and Georgia. As results, beginning from January of 2004 monthly expeditions were organized for the sampling in 35 points of Kura and Araks for determination of 12 heavy metals in water. Additionally, 2 pilot projects were organized in Azerbaijan to determine of concentration of main POPs compounds and radionuclides.

3. Nato SFP 977991 Project Results

3.1. HEAVY METALS

Investigation of 12 heavy metal was chosen for 9 transboundary and 3 internal sampling points in Azerbaijan (Table 1); 5 points for sampling waters after the dam: 3 in Kura tributary and Kura, and 2 in Araks and its tributary. The important sampling locations in Azerbaijan territory are:

- Mingechaur- After the dam, which captures the sedimentation process in this main water reservoir on Kura.
- Zardob- Situated approximately in the middle distance from Mingechaur till combining Kura and Araks, reflects heavy metal leaching and other activities in this agriculture region.
- One point after combining Kura and Araks. Samples from this point directly characterize the content of water which flow to the Caspian Sea.

Directly Araks water content after main influences from Armenia and Iran were represented in samples taken from Horadiz sampling point.

All samples were filtrated immediately after sampling through 0.46 filter and acidizing with HNO3.

TABLE 1. Sampling points in Azerbaijan territory

N-	River	Description of location	Latitude (in deg)	Longitude (in deg)
Az1	Chrami	Near Georgia boundary	41.3322	45.0686
Az2	Kura	After combining Chrami with Kura	41.2914	45.1830
Az-3	Kura tributary	AgstafaChay, after dam near Armenia boundary	41.0502	45.2714
Az4	Kura tributary	Iori, near Georgia boundary, after dam	41.1764	46.2473
Az5	Kura tributary	Alazan, between Georgian and Azerbaijan	41.2667	46.7083
Az6	Kura	After Mingechaur dam	40.7781	47.0349
Az7	Kura	Zardob city (center of Kura pathway between dam and Araks)	40.2069	47.7153

(Continued)

Az8	Araks	At Turkey boundary, near Armenia	39.6556	44.8035
Az9	Araks tributary	ArpaChay, After dam near Armenian boundary	39.6454	45.0857
Az10	Araks	After Nachchevan dam, between Iran and Azerbaijan	39.0853	45.4014
Az11	Araks	Horadiz settlement, between Iran and Azerbaijan	39.4414	47.3521
Az12	Kura	After combining with Araks	40.0724	48.5313

Table 2 shows of the main parameter for metals and observed ranges during 2 year monitoring.

TABLE 2. Data for investigated metals

		Crust	Oceanic	EPA drinking water standard	Minimal monitored value	Maximal monitored value
		mg/kg	ug/L	ug/L		
Silver	Ag	0.075	0.04	100	<0.1	0.56
Arsenic	As	1.8	3000	50	<0.5	17.6
Cadmium	Cd	0.15	0.11	5	<0.05	0.33
Cobalt	Co	25	0.02	50	<0.7	<0.7
Chromium	Cr	102	0.3	50	0.05	4.33
Copper	Cu	0.6	0.25	1000	0.78	18.63
Mercury	Hg	0.085	0.03	20	<0.6	1.5
Manganese	Mn	950	0.2	50	0.32	68.6
Molybdenum	Mo	1.2	10	40	0.8	16.2
Nickel	Ni	84	0.56	100	<0.7	8.93
Lead	Pb	14	0.03	15	<0.7	2.03
Zinc	Zn	70	4.9	5000	2.06	37.1

The analysis of data from Table 2 allows to conclude the following:

- Heavy metals content do not exceed international standards during monitoring processes. However, according to the FSU standards for Cd and Hg, which below 10 times the international standards, we registered elevated concentrations in some transboundary sampling points for both of these metals.

- More wide spans we have discovered for As and Mn. Most of elevated concentration of As were registered in Araks River- with mean value 11.82 ug/L, in the meantime mean concentration of As in Kura <1 ug/L.

Thus, we did not discover "expected" elevated concentrations for targeted hazardous compounds. These expectations were based on possible environmental contamination resulted from functioning of Armenian and Georgian mine industry, especially Cu and Mo production, developed during the Soviet period. These understated values may be explained by large water volumes in rivers during the last 2 years and low levels of mining production of metals in Armenia and Georgia during last years.

3.2. RADIONUCLIDES

Radionuclides concentrations were evaluated for both bottom sediment and soil samples taken adjacent to river sampling points.

TABLE 3. Sampling points for radionuclides

Description	Short label	Latitude	Longitude
Chrami (Kura tributary, which waters form in Armenia and Georgia) between Azerbaijan and Georgia boundary	Chrami	41.3188°	45.1188°
AgstafaChay (Kura tributary, which waters form in Armenia) near boundary with Armenia	AgstafaChay	41.0592°	45.2735°
Iori (Kura tributary, which waters form in Georgia) near boundary with Georgia	Iori	41.1758°	46.2477°
Kura between Yenikand and Mingechaur dams	Kura–Yenukand	40.9094°	46.3171°
Kura after Mingachevir Dam	Kura–Mingechaur	40.7855°	47.0309°
Kura till conjunction with Araks	Kura till Araks	40.0244°	48.4350°
Araz till conjunction with Kura	Araks till Kura	40.0096°	48.4298°
Kura after conjunction with Araks	Kura after Araks	40.0739°	48.5310°
Kura till 10 km to Caspian Sea	Kura at Caspian Sea	39.4150°	49.2651°

Samples were delivered after homogenization and storage for 1 month for equilibrium (in Marinelly beakers) and gamma-spectrometry analyses (Ge high resolution gamma spectrometer).

TABLE 4. Radionuclides in bottom sediment samples

	Ra-226	Ra-228	K-40	Co-60	Cs-134	Cs-137
	Bq/kg	Bq/kg	Bq/kg	Bq/kg	Bq/kg	Bq/kg
Chrami	20.1	25.5	471.1	<1	<1.1	9.6
AgstafaChay	12.9	15.6	431.3	<0.7	<0.9	0.3
Iori	33.5	36.9	625.2	<1.2	<1.4	2.2
Kura–Yenukand	53.0	61.8	1788.0	<3.3	<3.1	2.9
Kura–Mingechaur	24.1	27.2	569.2	<1.1	<1.3	1.5
Kura till Araks	36.2	42.1	728.9	<1.2	<1.4	1.5
Araks till Kura	35.1	34.2	717.8	<1.2	<1.4	2.6
Kura after Araks	33.3	35.5	755.6	<1.3	<1.4	1.1
Kura at Caspian Sea	33.5	41.9	743.9	<1.2	<1.4	2.1

As indicated in Table 4, activity of ^{60}Co and ^{134}Cs and were below MDA (minimal detectable activity). But ^{137}Cs values were registered for each sample, and there is obvious elevation of ^{137}Cs concentration in samples taken from Chrami bottom sediment near of Georgia boundary. By reviewing the Chernobyl catastrophe atmospheric deposition data we found information related to elevated precipitation of atmospheric isotopes in the region of Georgia during 1986.

Elevated concentration of ^{40}K for samples taken after Yenikand Dam on Kura relate with salt matrix of soils in this area. And ^{228}Ra (61.8 Bq/kg) and ^{226}Ra (53 Bq/kg) also will be related with salty sediment matrix which resulted by increasing of Ra dissolution.

3.3. PERSISTENT ORGANIC POLLUTANTS

Investigation of persistent organic pollutants was intended to estimate pesticides, poliaromatics and PCBs transport to Caspian Sea. Since all these compounds are poorly soluble in water, there was expected transport of these compounds via suspended matter. The sampling involves the collection of >50 L of water and trapping of suspended particle by filtration through a 0.45 um filter. The sampling points were determined to separate the Araks and Kura affects into the total transport. (Table 4).

TABLE 3. Sediment trapping points for organic pollutants (POPs)

Station Names	Latitude	Longitude
Kura Till– Sugovushan	40.019484°	48.446610°
Araz–Till Sugovishan	40.008969°	48.438961°
Kura After Sugovushan	40.024404°	48.460518°
Kura– Neftchala	39.416817°	49.244781°
Ana–Kur	39.357844°	49.352118°
Bala–Kur	39.369534°	49.349489°

Samples were delivered to the lab to investigated ranges of components. Results are given in Table 5.

TABLE 4. POPs compounds ranges

PAH		Pesticides		PCBs	
Name	Ng/L	Name	ng/L	Name	ng/L
Naphthalene	0.4–6.6	a-BHC	0.1–0.2	PCB-18	0.1–5.2
Acenaphthene+Fluorene	0.4–0.5	b-BHC	0.2–1.4	PCB-31	0.1–0.2
Phenanthrene	0.2–5.6	d-BHC	0.2–0.3	PCB-52	0.3–2.2
Anthracene	0.1–0.2	g-BHC	<0.1	PCB-44	<0.05
Fluoranthene	0.1–2.1	p.p'-DDD	2.1–7.4	PCB-101	0.4–1.1
Pyrene	0.2–0.8	p.p'-DDE	<0.23	PCB-149	0.5–2.6
Benzo(a)anthracene	0.2–2.0	p.p'-DDT	2.5–3.5	PCB-153	<0.06
Chrysene	0.2–3.0	Endosulf-I	0.1–0.4	PCB-138	0.1–5.7
Benzo(b)fluoranthene	0.2–1.8	Endosulf-II	0.4–0.8	PCB-180	0.1–1.6
Benzo(k)fluoranthene	0.1–0.4	Endosulf-sul	0.4–7.7	PCB-194	0.1–1.7
Benzo(a)pyrene	0.2–0.4				
Dibenzo(a,h)anthracene	1.2–17				
Benzo(ghi)perylene	0.3–1.2				
Indeno(1,2,3-cd)pyrene	<0.07				

There were detected p.p'-DDT and its derivate product p.p'-DDD in each of sediment samples. These were expected results because DDT was one of the intensive used pesticides in region during more than 50 years. Also Endosulfan–sulphate was registered in elevated concentrations. In list of PAHs were detected of Naphthalene, Phenantrene, Chrysene and Dibenzo (a, h) anthracene in elevated concentrations. Although all POPs concentration were below the "action level", there is a concern related to continuing transport of POPs from polluted agriculture and energy sector area to river water by storm waters and probably by ground water streams.

4. Importance of Reliable Data for Water Management

The collection of data of known quality for the main hazardous compounds was part of the new approach for water quality management:

- It is well known that 80% of Georgian sewage and 100% of Armenian sewage goes to Azerbaijan by river waters. The absence of industrial activity in these countries during the last years has resulted in dramatically decreasing of industrial sewage and river waters may have the lowest level of pollutants in compare with the 1980th. The detected low concentrations can be explained by strong decrease of impacts of artificial effects. However, one can expect of the increase levels of pollutants as Georgian and Armenian industry will be growing in the near future.

- It was discovered that a slightly increase in heavy metals content in Araks water during 2005 in comparison with 2004. This result may be related to some renewal of the mining industry in Armenia. Comparison of the collected data with future data will allow to separate the natural from artificial impacts into water content.

- Development of Ra Isotopes fractionation study with accompanying of Sr and Rn Isotopes investigations can provide data to derive information concerning ground water – surface water compartments in Kura–Araks waters along the rivers.

- Created database can be used for reliable modeling. Modeling will allow uncovering the full pattern of water quantity and water quality parameters.

- Water quality management can then receive a scientific basis for planning and implementation both for municipal, agricultural and industrial usage.

The Kura–Araks SfP 977991 project added two new components in its history. The first added the monitoring radionuclides, and the second added the study of POPs compounds. The project methodology allowed us to collect full scale data for modeling of transport and accumulation of hazardous compounds

Three new directions of studies can be proposed based on the obtained results:

1. More deep scientific research in the fields of environmental chemistry and environmental geochemistry. It is obvious that this topic should be addressed on a regional basis for South Caucasus than for each of the countries. As special subthemes the study of factors which controls absorption, sedimentation and leaching processes in specific sediments and water contents in regions could be conducted.

2. Development of the modeling component of the SfP 977991 project. The data collected within the project allow deriving some conclusion and forecasting on water quality contents already. The establishment of an international modeling team for Kura–Araks basin could generate new level of solutions both for local, national and regional water management practice.

3. Experience developed within the Kura–Araks SfP 977991 project could be applied in regional projects in other part of the world or in new projects involving of South Caucasus region.

References

Suleymanov B., Mansimov M., Ahmedov M. Harmonization of Azerbaijan surface water monitoring methodology– In "Problems of river monitoring and ecological safety of South Caucasus", CRDF, ISMCS program, TSU press, 2005

Durum W.H., Heidel G. and Tison L.J. (1960) Worldwide runoff of dissolved solids. International Association for Hydrological Science, 51, pp. 618–628.

Livingston D.A. (1963) Chemical composition of rivers and lakes. G. Data of geochemistry. United States Geological Survey Professional Papers, 440G, 1–64.

Tanabe S., Pourkazemi M., Aubrey D.G. –Concentrations of trace elements in muscle of sturgeons in the Caspian Sea Agusa T., Kunito T., Marine Pollution Bulletin, Nov 2004

Mora S., Sheikholeslami M.R., Wyse E., Azemard, S., Cassi R. An assessment of metal contamination in coastal sediments of the Caspian Sea Marine Pollution Bulletin, Jan 2004

Suleymanov B., Kerimov M., Determination of radioactivity and heavy metals in water samples taken from Kura River– Tans. of Azerbaijan National Academy of Science, 2000, 20, n1

Ibadov N.A., Huseynov V.I., Suleymanov B.A. Determination of polynuclear aromatic hydrocarbons by high performance liquid chromatography – Journal of chemical problems (Baku, Azerbaijan), 2004, N2, pp. 40–47

ON DEVELOPMENT OF GIS-BASED DRINKING WATER QUALITY ASSESSMENT TOOL FOR THE ARAL SEA AREA

DILOROM FAYZIEVA
Institute of Water Problems, Academy of Sciences of Uzbekistan

ELENA KAMILOVA
Institute of Genetics and Plant Experimental Biology, Academy of Sciences of Uzbekistan

BAKHTIYAR NURTAEV
Institute of Geology and Geophysics, Academy of Sciences of Uzbekistan

Abstract: This paper presents and discusses results of the GIS-based studies of the Zarafshan River including the territories of the Samarkand, Navoi and Bukhara provinces, where the epidemic situation from waterborne infections occurs, especially for typhoid fever and bacterial dysentery. Application of GIS allows complex studies to be conducted on the epidemiological conditions of this area which are necessary for monitoring, analysis, and prognosis of water supply sources and assessment of their influence on population health as well as for the decision-making process of the sanitation and epidemiology centers and other environmental protection organizations.

Keywords: transboundary rivers, public health, waterborne diseases, potable water sources, Zeravshan River, Aral Sea

1. Introduction

The present environmental situation in Central Asia is characterized by deterioration of sanitary and epidemiological conditions in the basins of transboundary rivers. It is caused by a practice of unregulated consumption and use of water by different water consumers which have common areas of water consumption. This situation requires a comprehensive study of these water bodies aimed not only at determination of their condition at a given time period but also at forecasting their sanitary condition by studying conditions of water consumption and the subsequent effects of water quality on human health.

The problem of safe potable water supply to population in Uzbekistan is challenging because of the growing water deficit and difficulty of systematic monitoring of water quality. To improve the management of water quality, new information technologies and advanced concepts of evaluation of risk for population health need to be introduced [13]. The application of new information technologies makes it possible to analyze, control and forecast the sanitary conditions of the bodies of water that are used as the sources of water supply and, at the same time, are affected by the anthropogenic factors originated by irrigation, washing out fields, industrial and domestic wastes [1,6,14]. The quality of evaluation of environmental situation to a great extent depends on availability of information technologies allowing managers to utilize large databases [2,6].

A solution to the problem of providing the population with safe potable water in the region is becoming more urgent because of the increasing deficit of water resources and difficulties in systematic monitoring of water quality. To improve the practice of drinking water quality management and prevention of negative consequences on population health it is necessary to implement new information technologies. The major objectives of the present study are to improve the methodology of evaluation of impact of water quality on public health using geographical information systems (GIS) including collection, systematization of the existing data (formation of the data base); to perform the information analysis on water indicators of the water reservoirs that are affecting by various kinds of contamination (microbiological and chemical); and to evaluate possible effects on the health indicators.

The research on GIS-based assessment of waterborne diseases was carried out in the framework of the Science and Technology Center of Ukraine (STCU) Project No 3173, "Development of new information tools for determination of water quality in the Aral Sea area".

The major goals of this research are to:
- Collect, systematize, and analyze the information on evaluation of potable water quality and the dynamics of health indicators, i.e., to create a database.
- Apply the information geographic system (GIS) to improve the methods of spatial analysis of water related health status of population.

To achieve these goals, the study on waterborne diseases assessment using GIS was divided into the following parts:
- Studying the factors that form the sanitary-bacteriological status of potable water sources.
- Analyzing the health status of population by the morbidity statistics of intestinal infections, the spread of which, no doubt, is determined by water factors.

2. Materials and Methods

The study area was the basin of the River Zarafshan including the territories of the Samarkand, Navoi and Bukhara provinces, where the epidemic situation from waterborne infections occurs, especially for typhoid fever and bacterial dysentery.

The study was based on the ecological and epidemiological data obtained from organizations responsible for environmental protection and public health. The spatial analysis was performed using the software MAPINFO which allowed us to utilize a geographical data base and to reveal the most unfavorable areas of the investigating territory from the point of view of population morbidity with intestinal infections. The information on bacteriological indicators of water quality had been summarized for the period from 1996 till 2005. The data on morbidity rate of the population with waterborne diseases (typhoid fever, bacterial dysentery, other diarrhea diseases and viral hepatitis A) for that period had been also investigated in the territory. Thematic maps were created on the basis of ecological and epidemiological data of different territories and it was found that zones with high rates of morbidity tend to be associated with unfavorable water conditions.

The studies of physical-geographic and other natural factors influencing the condition of water-bodies in the Zarafshan Basin and evaluation of anthropogenic factors that might impact on water quality in water-bodies being used for potable water supply were based on the documents of some agencies and institutions dealing with water quality monitoring, e.g., the Goscompriroda (Nature Protection Committee), Glavhydromet (Main Unit for Hydrology and Meteorology), and the Center for Sanitation and Epidemiology of the Republic of Uzbekistan.

The territory of the Samarkand region was chosen for the research using the GIS. The data on population morbidity were taken from the department of the State Sanitary and Epidemiology Surveillance, Statistic and of the Health Ministry, regional health administrations and regional sanitary and epidemiology surveillance administrations of the Samarkand region of Uzbekistan. The samples of surface and drinking water were taken and analyzed for bacteriological indicators. Analysis of drinking water quality was carried out in two large cities (Samarkand and Kattakurgan), and in district centers of Samarkand region from existing sources of water supply (water system, river and canal water, and wells). Water samples were taken and analyzed according to State Standards [15].

The spatial analysis was made using the GIS MapView that enabled us to utilize the geographical database and to reveal the most affected areas in view of the acute intestinal diseases rate among the population. The map (of 1:500,000 scale) was one of the basis for the GIS.

3. Results and Discussions

3.1. EVALUATION OF CHANGES IN THE QUALITY OF WATER IN THE RIVER ZARAFSHAN

The River Zarafshan is in Amydarya River Basin. The Zarafshan's basin is characterized by an extremely high number of pollutants that penetrate into it from point and diffuse sources of pollution. The Zarafshan is considered to be one of the most contaminated rivers in Central Asia. This is caused by a high density of population and a large number of industrial and agricultural discharges. The Zarafshan River starts in the territory of Tajikistan where its water is intensively used and enters into Uzbekistan in a polluted state. Within Uzbekistan untreated waste water is discharged into the Zarafshan by 2,500 industrial enterprises located in central areas of the country. The severest pollution with agricultural drainage water was observed in the lower reaches of the Zarafshan. The diffuse sources of pollution include: heavy rains drained from built-up areas and waste water from the land involved in agricultural production.

The river is the main source of potable water supply to the local population. In the basin a tense situation with water supply and environment is developing because the social and economic needs of the region require a significant increase of water supply for developing industry, household services and irrigated crops. This situation manifests itself in the sanitary and epidemiological changes as most of inhabitants of the rural area are using water for drinking and household needs from surface water-bodies and canals. The qualitative and quantitative shifts in microbial cenosis have been registered resulting from poor flow caused by huge water withdrawal for irrigation and intensive pollution of water with organic and biogenic substances. In the Samarkand region, the phenol concentration, (i.e., the main indicator of pollution of water-bodies with organic substances), reached 0.02mg/l (20 MPC) while the maximum permissible concentration is 0.001 mg/l. Such concentrations of phenol are caused not only by anthropogenic pollution but also by the natural features of the region.

Dissolved oxygen, carbon dioxide, nitrogen compounds, soluble phosphorus, and total iron are the most important substances that influence the metabolism and catabolism of water microorganisms. The BOD characterizes the level of organic pollution of water-flow, and its value grows with an increase of anthropogenic impact. High BOD values were found almost everywhere, in the basin, even during higher flows in the Amankutan, Sazagan, and Tusunsai. BOD value was found at concentrations of 4.16 mg/l which is many times higher than the MPC. The quality of potable water in the basin of the river Zarafshan has shown a tendency to deteriorate over time. Today a serious danger for human health in the region is constituted by poor provision of sewerage system services and lack of water-treatment plants. The major part of the population lives in the rural area where water pipe systems are poorly developed.

Intensive biogenic and organic pollution of water results in misbalance events in the ratio of microorganism species in water-body causing an increase of the level of pathogenic bacteria (Salmonella) in the range of 16–33 MPN (most probable number) per one liter of water even when the content of the indicator microflora is decreased. Identification of pathogenic microorganisms in the water (Salmonella, Shigella, etc.) indicates potential epidemiologic danger of water reservoirs as the source of intestinal infections.

3.2. GIS-ANALYSIS OF THE RATES OF WATERBORNE ACUTE INTESTINAL INFECTIONS AMONG THE POPULATION OF THE SAMARKAND REGION

An important condition influencing the choice of an information technology in view of the goals and objectives of the population research are space-time characteristics that are necessary to analyze various data on environment and population health [3,4,5,7]. Evaluation of medical and environmental factors in time and space should be done using the standard procedures enabling their comparison with various data on the place, environment status and population. Geographic information systems (GIS) may be successfully used for these purposes. This information technology is accepted as a standard approach to be used for analysis of spatial data and is one of the best both for collection and processing indicators of environment status and evaluation and analysis of its impact on human health. The GIS efficacy is mainly based on the fact that the same system of coordination of data is used for many fields of knowledge that makes it possible to compare diverse information [2,8,9,10].

Among the most common infections in Uzbekistan are salmonellosis, shigellosis and other acute intestinal infections, i.e., acute diarrhea diseases. From an epidemiological point of view, the most significant waterborne infections are typhoid fever and bacillary dysentery.

For the last decade the Republic of Uzbekistan had been one of the "leaders" among the CIS members in typhoid fever rate. At present, despite a significant decrease of typhoid fever rate (for the last 13 years it has decreased by 11.3 times), it is still rather high being 7 times higher than the rate in Russia: 1.5 versus 0.22 per 100,000 population [11,12].

A characteristic feature of epidemic process of typhoid fever, that is a typical representative of intestinal infections, is the group (outbreak) morbidity. Taking into account a high proportion of waterborne outbreaks, composing 54.2% of the total number of diseases outbreaks, the researchers assert typhoid fever to be mainly transmitted in Uzbekistan through water [11,12]. An intensive growth of the morbidity rate is registered in hot seasons when water consumption increases and surface water-bodies are used for recreation.

The information for the five years (1994–1998) obtained for each indicator to be predicted was summarized. In addition, the data for 1996–2004 on morbidity of typhoid fever and bacterial dysentery among the population in the area under study was analyzed.

A territory of a district was taken as a unit of taxonomic characteristic that is convenient in terms of administrative relations and statistical data. However, these areas do not correspond to the areas of contamination with ecotones and actual landscape and geographic borders. This required the creation of specific maps based on topographic and landscape maps.

Figures 1 and 2 present the information on the rates of typhoid fever and bacillary dysentery. These figures illustrate the average values obtained during a long-term observation (1996–2004) conducted in different districts of the Samarkand region. These data were obtained during a comprehensive surveillance of the sanitary and epidemiology status of the region. The research findings show a marked discrepancy of spatial distribution of typhoid fever and dysentery cases that depends on the hyper-endemic character of some areas in the Bulungur and Urgut districts of the studied region.

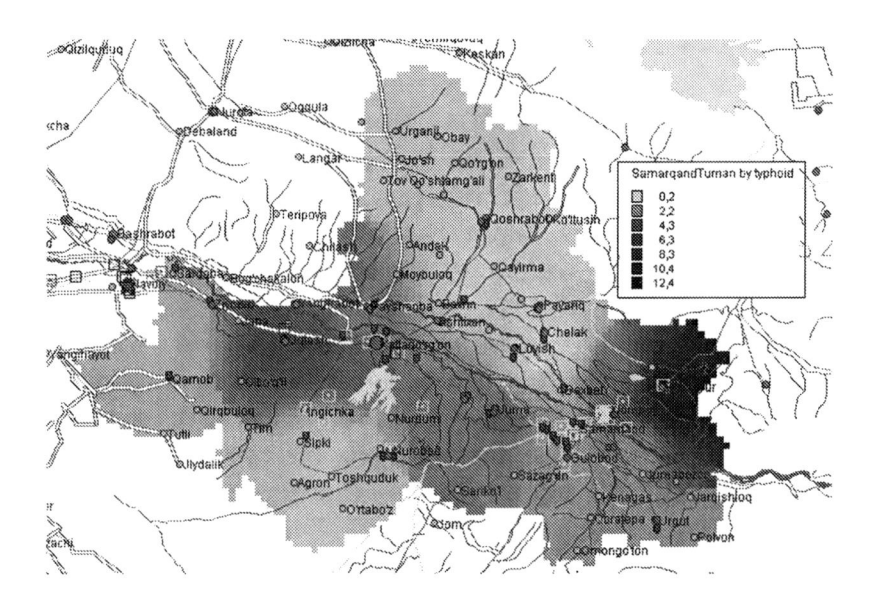

Figure 1. Typhoid fever morbidity among the population (the number of cases per 100,000 population)

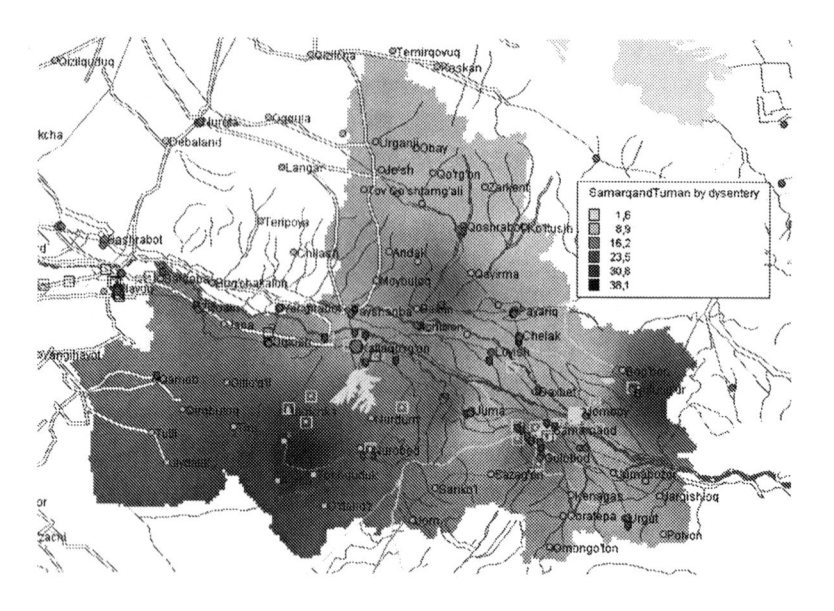

Figure 2. Distribution of dysentery morbidity among the population (the number of cases per 100,000 population)

This fact once more suggests a differential approach to evaluation of the situation and further revealing the conditions and agents of the epidemic process activation. The epidemic surveillance needs to be improved to address the epidemiological situation at definite time and place.

Recently the unfavorable dynamics of bacteriological indicators of water quality had been registered in the places of water intake for potable water supply in the Samarkand region. In 1996 the sanitary standard was not met by 14.5% of water samples taken in surface water-bodies, while in 2004 the figure increased to 44.5%. This indicates an increased microbe pollution in water-bodies and a higher risk of intestinal diseases morbidity for the local population. Taking into account the information on availability of centralized water supply in the urban (83.5%) and rural areas (71.5%) of the region, one can assume how much the rural inhabitants are likely to get intestinal infections. The rates of waterborne infections in rural areas were significantly higher in comparison to the urban one. This assumption can be confirmed by the fact that most of patients with typhoid fever and dysentery (97.9% and 93.5%, respectively) in the Samarkand region were people living in the rural area. This situation is probably also caused by the character of their jobs as most of them are working in agriculture and have frequent contact with water and soil. Revealing the cause–effect relation of these diseases development is a rather difficult process as the impact of other environment factors on disease development is complex and requires further in-depth ecological and epidemiological research before development of sound preventive actions in the region.

4. Conclusions

The approved methods of the epidemiological analysis using the geographic information system helps with the evaluation of environment risk for human health. The GIS technology ensures joining up the materials of different agencies, organizations and institutions of industrial, nature protection and medical profiles and enabled us to form a holistic conception on the risk of getting infected with waterborne diseases in a definite area.

This technology needs to be used for monitoring and prediction of water-bodies status and evaluation of the impact on human health that are necessary both for the practice of sanitary-epidemiological service and departments of environment expert examination in nature protecttion agencies. The same approach needs to be approved in other regions of Uzbekistan where the situation with water supply is also in short supply. Application of GIS allows carrying out complex studies on epidemiological conditions of this area and it is necessary for monitoring, analysis, and prognosis of water supply sources and assessment of influence on population health as well as for the decision-making practice of the sanitation and epidemiology centers and other environmental protection organizations.

Acknowledgements

This collaborative study involving researchers from three institutes of Uzbekistan Academy of Sciences is being completed in the frame of the project No 3173 "Development of new information tools for determination of water quality in the Aral Sea area" funded by the Scientific and Technology Center of Ukraine (STCU), 2004–2007. We wish to thank STCU for their funding and support of this project. We are indebted to Dr. Danny D. Reible, from the University of Texas at Austin for his continuous readiness to assist us with scientific and personal advice.

References

Aral M.M., and Maslia M.L., Evaluation of human exposure to contaminated water supplies using GIS and modeling. In: Kovar K and Nachtnebel H.P. (Eds.) Application of Geographic Information Systems in Hydrology and Water Resources Management. IAHS Publications No 235, pp. 243–252, IAHS Press, Wallingford, 1996.

Barr R. Data, information and knowledge in GIS. GIS Europe., 5. No 3, 1996.

Environmental Information Systems in the Russia Federation an OECD Assessment. Paris: Organization for Economic Cooperation and Development, 1996,106 pp.

Frisch J.D., Shaw G.M., Harris J.A., Epidemiologic Research Using Existing Databases of Environmental Measures. Archives of Environmental Health, 1990,45, No 5, pp. 303–307.

Geographical & Environmental Epidemiology. Methods for Small Area Studies, P.Elliott, J. Cuzick, D. English and R.Stern, Oxford Medical Publications, 1996, 382 pp.

Malone J.B. et al. Use of LANDSTAT MSS imagery and soil type in a geographic information system to assess site-specific risk of fascioliasis on Red River farms in Louisiana ,Annals of the New York Academy of Sciences,1992, 653, 389–397.

The Added Value of Geographical Information Systems in Public and Environmental Health. Ed. Marion J.C. de Lepper, Henk J. Scholten and Richard M. Stern,, Kluwer Academic Publishers, The Netherlands, 1995, 355 p.

Understanding GIS: The ARC/INFO Method. Environmental System,Research Institute Inc. 1990, Apple.

Application of GIS-technologies in organization of the data in big corporations, ARC REVIEW No2. 1999, p. 3.

Martynenko A.I., Bugaevsky U.L., Shibalov S.N. GIS fundamentals: theory and practice, Moscow, 1995 p. 75.

Mirtazaev O.M., Norov G., Hasanov A.H., Toshpulatov E., Tursunov P.B. Epidemiological characteristics of typhoid fever outbreaks, Infections, Immunity and Pharmacology. 1999. No 1, pp. 78–80.

Mirtazaev O.M., Niyazmatov B.I., Norov G., Kantemirov M.P, Epidemiological characteristics of typhoid fever in Uzbekistan, Epidemiology and Infectious Diseases, 199, No 4, pp. 16–18.

National Action Plan on Hygiene of Environment of the Republic of Uzbekistan, Tashkent, 1999.

Reshetov V.V., Gorky A.V., Comprehensive evaluation of radiation, geo-chemical, medical and sanitary situation in towns of Russia Scientific report on implementation of the 2nd stage of the Federal Program,"Ecological Safety of Russia", 1994.

RST Uz 951: 2000 (the State Standard) "Sources of centralized industrial and potable water supply", Tashkent, 2000;

Utkin V.V., Problems of formation of the territory information space // Information Technologies, 1996, No 3, pp. 16–19.

PART III: LEGAL, TECHNICAL AND INSTITUTIONAL ASPECTS OF TRANSBOUNDARY WATER MANAGEMENT

LESSONS LEARNED FROM TRANSBOUNDARY MANAGEMENT EFFORTS IN THE APALACHICOLA–CHATTAHOOCHEE–FLINT BASIN, USA

STEVE LEITMAN
Institute for International Cooperative Environmental Research, Florida State University, Tallahassee, Florida, USA

Abstract: This paper serves as an example of transboundary water negotiations – what worked and what failed – in the three state conflict in the southeastern United States. The states of Alabama, Florida, and Georgia have been involved in water negotiations since about the early 1970s leading to some successes and some failures. This paper summarizes these experiences with the intention of providing generic "lessons learned" on multi-jurisdictional negotiation.

Keywords: southeastern United States, lessons learned, policy/political/technical aspects of negotiation

1. Introduction

The Apalachicola–Chattahoochee–Flint (ACF) drainage basin is a 50,000 km^2 basin located in the southeastern United States (Fig. 1). Average flow at the mouth of the watershed is about 700 m^3/s. Figure 2 shows the variation of average monthly flow over the course of the year. The waters of the ACF basin are used and managed for multiple purposes including water supply, waste water dilution, hydropower production, commercial navigation, recreation, flood control, and sustaining and harvesting natural resources.

Only the Chattahoochee River has the capacity to regulate flow through storage reservoirs, although there is a reservoir with very limited storage at confluence of the Flint and Chattahoochee rivers. Although some of the reservoirs in the basin are operated by private interests such as the Southern Company, virtually all of the reservoir storage capacity is at federally operated reservoirs. Because of the limited storage capacity relative to flow in the basin and the fact that about 2/3 of the storage capacity of the river is located in the upper Chattahoochee Basin above metropolitan Atlanta, the ability to store flood waters and augment flows during periods of drought and low flows is relatively limited in comparison with many other watersheds in the United States. Weighted average rainfall in the basin is about 135cm/year and rainfall tends to be greatest in the winter and summer, and least in the fall.

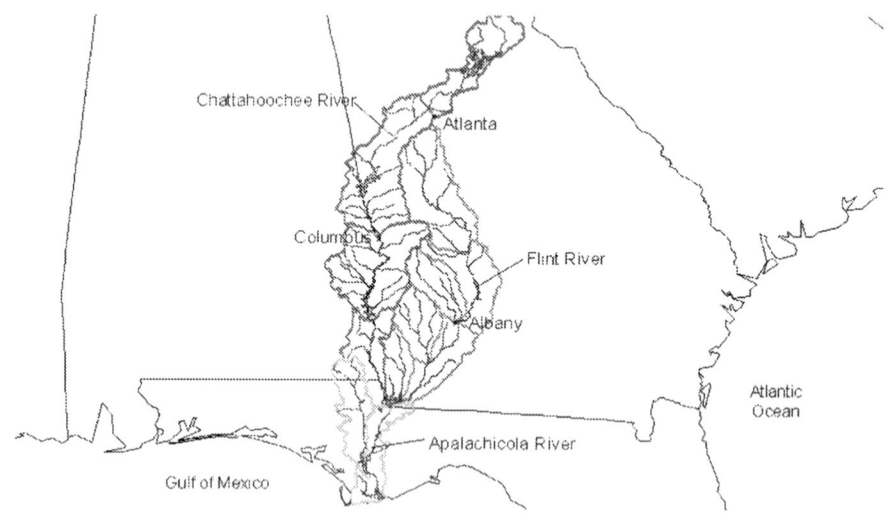

Figure 1. The ACF basin

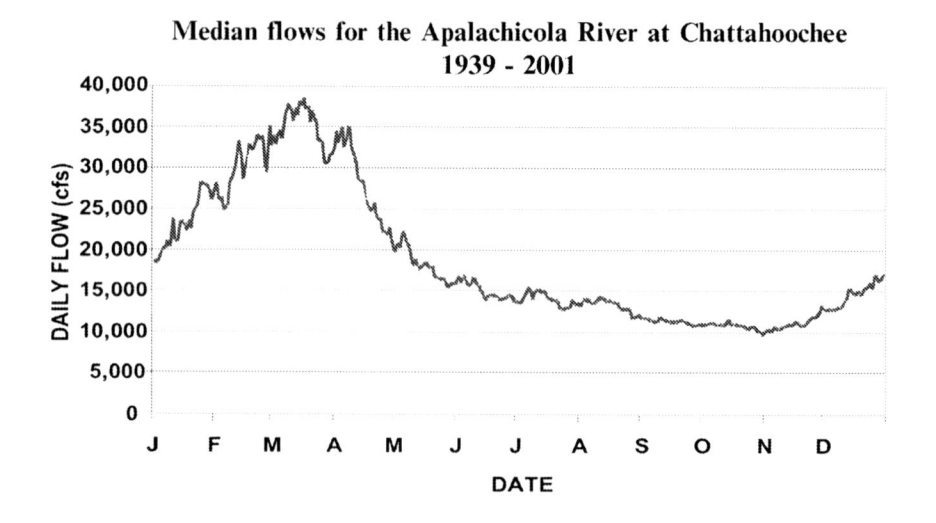

Figure 2. Median flows on the Apalachicola River

The major political entities in this watershed are the states of Alabama, Florida, and Georgia, the federal government and the Atlanta metropolitan area. About 75% of the basin lies in Georgia, 1/8 in Florida and 1/8 in Alabama, so consequently the flow in the downstream portion of the basin (Florida) is defined by rainfall patterns, usage and upstream management in the upstream portion of the basin.

The management goals of the major political entities vary in a predictable manner. As an upstream state, Georgia's management interests are based on maximizing withdrawals for users within the state and keeping storage reservoirs full to support withdrawals in periods of drought and provide for water-based recreation at other times. Georgia is also interested in hydropower production and commercial navigation (which is dependent upon channel depths in the Florida portion of the river).

The metropolitan Atlanta area is the largest metropolitan area in the basin and it wields enough power in both the State of Georgia and the southeast to be considered as a major political entity in this dispute. It should be noted that Atlanta holds the distinction of being one of the few major metropolitan areas in the United States which is located in the headwaters of a basin. Essentially this puts Atlanta in the same water supply situation as major metropolitan areas located in much drier areas of the United States such as Los Angels, Phoenix or Las Vegas rather than other metropolitan areas in regions with similar rainfall. Atlanta's interests in the basin are to support its ever growing demands for water since the Chattahoochee River represents the cheapest source of water for "Metro Atlanta" and to maintain the elevations of Lake Lanier, the largest reservoir in the basin and a major recreational area for Atlanta residents.

Alabama's water management goals for the ACF basin are more focused at preserving future options for water withdrawals in order to attract economic growth from the ACF basin than in securing water for an existing use. Alabama also has a long history of favoring management of the federal reservoir system to support having a commercial navigation channel in the Apalachicola River. Alabama's management goals for the basin are complicated by the fact that management of the adjacent basin, the Alabama–Coosa–Tallapoosa (ACT) basin (Fig. 3), has been linked with management of the ACF basin through lawsuits and negotiated agreements. Since far more of

Alabama is in the ACT basin than in the ACF basin, their negotiating position was more aimed at trading their influence in the ACF for protection of the ACT, than in specifically advocating for actions in the ACF basin. State officials were also concerned with long-term water quality problems resulting from discharges in the Atlanta region.

Florida's management goals have been focused on protecting the instream flow of the Apalachicola and the flow entering into the Apalachicola estuary. Florida has a long-term record of advocating for the protection of the Apalachicola Basin and over the past 30 years the state has purchased over 50,000 ha of land for conservation purposes, imposed every protective designation available, resisted the construction of a dam and other structural improvements for the federal navigation channel and supported extensive research to protect this ecosystem. Apalachicola Bay produces approximately 15% of the nation's oysters as well as extensive yields in shrimp, blue crab and finfish, and it serves as an important nursery grounds for the Gulf of Mexico. In contrast to Georgia's and Alabama's water needs, Florida's needs are not well defined since the science behind providing adequate inflow to protect an ecosystem is not well developed.

The Federal government's management interests in the ACF basin pertain to legislated responsibilities that the federal government has for the management of the federal storage reservoirs for producing hydropower and supporting federally maintained navigation channels. The federal government also has natural resource oriented responsibilities, including protecting federally listed endangered species such as the gulf sturgeon and several species of mussels which live in the river. The federal government also has to manage the federal storage reservoirs to provide balance between adequate elevation in the reservoir to support reservoir-based recreation and making releases to support down steam flow needs. Lake Lanier (in the metro Atlanta area) has among the highest recreation visitation rates of the reservoirs in the United States.

2. Efforts to Manage the ACF Boundary from a System-Wide Perspective

After several failed efforts at initiating system-wide management of the water resources in the 1970s and 1980s, an attempt to initiate such a management approach was made in 1989. A contentious relationship among competing water users in the ACF basin extends back to the 1970s as a result of the limited availability of the federal navigation channel. Upstream interests contended that the limited availability of the navigation channel hindered their economic development, whereas Florida refused to allow the construction of major structural alterations to the Apalachicola River to address navigation problems because of associated adverse environmental effects and because most of the benefits from the project were to be accrued by upstream interests not within Florida. Without these changes, the ports on the ACF river system, which are mostly in Alabama and Georgia, have limited access to the rest of inland navigation system in the United States.

At the time, the argument was seen as a conflict between environmental interests not allowing the complete structural modification of the basin and navigation interests desiring a more reliable channel. It was not until over 15 years later that government entities throughout the basin began to accept that the true conflict was over the amount of water available for all uses, not over the obstructionist tactics of environmental interests. The reason for this delay is probably the fact that the basin lies in a relatively humid region and a long prevailing attitude that scarcity of water was not seen as an issue in the region. Water managers tended to see the problem as a management problem, not a supply problem. It was convenient to blame another party rather than accept that they were pressing the system's limits during low water events.

In 1989, the Corps of Engineers proposed to reallocate water in the storage pool of Lake Lanier from hydropower to water supply for the metro Atlanta area and to formalize current reservoir operations in the form of a Water Control Plan (USACE 1989). Upstream interests reacted by contending that the federal reservoirs were being used too much to support downstream needs and downstream interests reacted by contending that too much water was being consumed and retained upstream. As a result of including reservoir operations with the reallocation proposal, attention expanded to the entire watershed instead of just the headwaters of the Chattahoochee Basin. There was a widespread fear that Atlanta's water use would dry up the river, a fear that persists to this day.

In response to this proposal, Alabama sued the Corps of Engineers for failing to meet the requirements of the National Environmental Policy Act in their preparation of the required Environmental Impact

Statement. With Florida poised to enter the suit on the side of Alabama and Georgia on the side of the Corps of Engineers, the three states and the federal government negotiated an agreement to stay the suit and conduct the ACF Comprehensive Water Resources Study. The Comprehensive Study provided technical information, developed tools to evaluate water resources from a system-wide basis and collected technical information on the management of river basins.

The Comprehensive Study, in turn, led to the establishment of the Apalachicola–Chattahoochee–Flint River Basin Compact in 1997. This Compact established the ACF River Basin Commission and required it to "establish and modify an allocation formula for apportioning the surface waters of the ACF basin." Establishment of this Compact and its sister Compact in the ACT basin was significant because it was the first such Compact ever in the southeastern United States and first in the nation since passage of major environmental legislation in the mid-1970s. It should also be recognized that in the United States the only means of addressing system-wide water management is either through establishing a Compact, federal legislation or through a decree by the US Supreme Court as a result of litigation (Leitman 2005; Dellapenna 2006).

The Compact did not include specific details of an Allocation Formula because the Comprehensive Study had demonstrated that other Allocation Formulas in the United States have needed to be changed over time and if the Formula had been included in the Compact, new legislation would have had to have been passed through the legislatures of the three states and the US Congress to make any changes. The legislative difficulty of changing other Compacts has resulted in litigation before the US Supreme Court (Kenney 1996). If the Allocation Formula was delegated to the control of ACF River Basin Commission, the formula could be amended without legislative approval.

The Allocation Formula was negotiated through the ACF Commission. This Commission consisted of the governors of the three states, although the actual negotiations were conducted by their appointed representatives, not by the governors themselves. If an agreement could be reached between the three states, the federal government had 245 days to either accept or reject the agreement. Ultimately after 14 extensions of the deadline for reaching an agreement on the Allocation Formula the negotiations were terminated in the fall of 2003. The termination of the negotiations can be attributed to a breakdown in trust among the negotiating parties (Dellapenna and others 2006). In the final meeting of the ACF Commission, Alabama, and Georgia were in favor of extending the negotiations again, but Florida refused to extend them and this meant the termination of the ACF Compact. It also meant a change in the forum for addressing water management issues from the negotiation table to the courtroom and a change in approach from collaboration to competition. The demise of the Compact can be blamed on both the process used to negotiate an agreement and the breakdown in trust among the negotiating parties due to action both in the negotiations and outside of the negotiations (Leitman 2005).

Among the problems with the process of the ACF Allocation Formula were:

- Not including an outside mediator to facilitate negotiations.
- Not agreeing on specific criteria or performance standards that distinguished an acceptable agreement from and an unacceptable one.
- Having parties involved in the negotiations enter into a Settlement Agreement on litigation involving issues that were part of the Allocation Formula negotiations.
- Having the governors of the three states define the terms of an acceptable agreement through a Memorandum of Understanding (MOU) after negotiations had been ongoing for five years without the involvement of all key parties and stakeholders in the negotiations.
- And, setting up a forum for negotiation which was not conducive to negotiating.

The MOU between the three governors in July 2003 led directly to the termination of the Compact. The MOU was negotiated between the governor's offices of the three states to provide the basis for negotiating an Allocation Formula agreement. The principal ACF negotiator for Florida, however, was not involved in developing this MOU and did not even see it until it was presented at the Commission meeting at which it was adopted. The MOU was essentially an endorsement of Georgia's negotiating position that Florida had rejected numerous times because it violated several of Florida's main negotiating positions.

After the agreement was signed, attempts were made to add stipulations to the Agreement to reaffirm some of the basic tenets of the Florida negotiating position. When Georgia's negotiators sent the MOU to their stakeholders, they did not include Florida's stipulations and contended that they never received them. Georgia then provided an Allocation Formula proposal to Florida that was consistent with the MOU, but unacceptable to Florida negotiators and stakeholders. In response, Florida provided an alternative proposal to Georgia one week before negotiations were to terminate that was consistent with their stipulations, but not consistent with the MOU. Florida negotiators told Georgia negotiators that if they did not accept the terms of their alternative proposal, Florida would not agree to extend the negotiations and the Compact would be terminated. At the final meeting, Alabama and Georgia expressed a desire to extend the negotiations and a disappointment that Florida was not following the terms of the MOU.

Although the termination of the ACF Compact suggests that this effort was a failure, there were several major gains to the citizens of the basin as a result of the Comprehensive Study and Allocation Formula negotiation process (Leitman 2005). Some of the gains from this effort included:

1. The paradigm for managing the basin for many stakeholders has expanded from a parochial or local perspective to a watershed perspective.
2. A significant amount of information, data and management tools were developed and are now available to address water management issues in the present and into the future.
3. There were multiple institutional changes in the three states and among nongovernmental organizations as a result of the negotiations.
4. A number of new management paradigms have become part of the "management vision" for the watershed including adaptive management, protection of flow regime versus sustaining minimum flows and the shared vision planning.

At the present time, the parties are maneuvering toward a Supreme Court challenge and the chances of rebuilding the trust "seem remote at best". In August of 2006 the parties were in dispute over the operations of the federal reservoirs and protection of the gulf sturgeon and several species of mussels that are protected species under The Endangered Species Act. The basin was experiencing a major drought event and the dispute was over whether water should be released from the federal reservoirs to protect the listed species at the present time or whether the water should be held in storage in case the drought event should persist. The reservoir augmentation needed to protect the listed species was to offset consumptive losses of water from municipal water users, agricultural water users and evaporation losses at reservoirs.

For the balance of this paper the focus will be on "lessons learned" from this attempt at system-wide water management and at the potential applicability of these lessons in Central Asia.

3. Lessons Learned from the ACF Compact Experience

LESSON 1: IN DEALING WITH TRANSBOUNDARY WATER ISSUES, PATIENCE IS A VIRTUE BECAUSE IT CAN TAKE A LONG TIME TO EFFECTIVELY ADDRESS A COMPLEX PROBLEM

Because of the complexity of the process of negotiating a transboundary water dispute and the fact that it normally takes many years to create many of the problems that lead to the dispute, it should be expected to take some time to successfully address the problem. History has shown that it is not uncommon for it to take 5 to 10 years or even longer to work out such problems (Wolf, 2001).

At the present time it is difficult to understand whether these efforts were successful or a failure. If an Allocation Formula had been agreed to which did not resolve the issues at hand had been agreed to, is this a success? For instance, an agreement was reached in the Colorado River, but this agreement resulted in an over-allocation of the waters of the basin. In the case of the ACF negotiations and litigation, although there was no agreement on a water allocation formula, the data and tools developed in the ACF Comprehensive Study/Allocation Formula negotiations have proven to be important in developing interim reservoir operations to protect endangered species. If an Allocation Formula had been agreed to that did not resolve the issue, would this really be a success?

LESSON 2: PROCESS IS AS IMPORTANT AS PRODUCT. PARTIES NEED TO BELIEVE THE PROCESS IS IMPARTIAL AND KEY PARTIES NEED TO BE PART OF THE SOLUTION

A major focus of alternative dispute resolution practices is to develop a process which is conducive to the parties reaching an agreement. In negotiating a long-term water agreement it is important that all parties and key stakeholders believe in both the agreement and the process under which it was developed if the terms of this agreement are to be sustainable. In the ACF negotiations an attempt was made to reach an "agreement" by having the governors of the three states develop an MOU to define an acceptable agreement without involving neither key personnel in the negotiations nor key stakeholder groups in an attempt to resolve the dispute. This tactic, however, did not resolve the problem but perhaps was the final nail in the coffin that led to the demise of the Compact agreement.

Another lesson from the ACF negotiations with regard to process is that negotiation is not always the best process to resolve such disputes. Negotiation only works when all parties are serious about negotiating and willing to focus on interest-based negotiations. All parties must have more to gain from a negotiated agreement than they are willing to give up or else they will not be negotiating in earnest. Negotiation then becomes a tactic to get what a party wanted all along, not a means to address each party's legitimate interests. In such cases it may be necessary to proceed to litigation until all parties are serious about negotiating, which may never occur.

LESSON 3: THE FOCUS NEEDS TO BE ON GETTING IT RIGHT, NOT ON BEING RIGHT. SUSTAINABLE ANSWERS MUST BE FLEXIBLE AND ALLOW FOR LEARNING AND REVISIONS OVER TIME. FIXED ANSWERS WILL BECOME STALE OVER TIME

The focus of negotiations needs to be on generating the necessary information and tools to addressing the complex problems at hand, not finding a politically expedient manner to pass the problems on to future generations. A process needs to be agreed to and implemented that will allow the parties to address contentious and difficult issues in an objective manner. For this reason, I believe it is necessary to have outside technical parties who have no stake in the results and a mediation team which will keep the process on task and objective. In the ACF negotiations there was no outside mediation team and this ultimately led to an avoidance of dealing with difficult technical issues by putting them off and then extending deadlines over and over again. The process also needs to account for the fact that the technical community probably cannot answer all questions at the present time. This leaves the parties with several choices: pretend they know the answer to all difficult questions, ignore questions they cannot address or set up an adaptive process which allows for learning while the agreement is being implemented and modifying the agreement to include what is learned. A problem with modifying agreements which needs to be accounted for *a priori* is that modifications will inevitably favor one party over another and if implementation is left to a consensual process, the party that is not being favored can be expected to oppose such a modification.

LESSON 4: IT IS IMPORTANT TO HAVE JOINT TOOLS TO APPROACH THE PROBLEM. MODELS ARE ALWAYS PART OF MAKING COMPLEX DECISIONS, THE ONLY QUESTION IS WHETHER MODELS ARE COVERTLY IN SOMEONE'S HEAD OR OVERTLY DOWN ON PAPER OR IN A COMPUTER

It is not uncommon for technical teams to get caught up in arguments over whose modeling tool is better or whose data are more accurate. One way to avoid these arguments is to initiate the negotiation process by having the parties jointly gather data and develop shared tools to analyze the data such as is done through the shared vision process (Palmer and others 1999, Stephenson 2001). The process of gathering data and developing tools to accurate represent the watershed allow the parties to develop trust in working together before having to tackle the more difficult and contentious problems which will inevitably result from defining an acceptable course for sharing the waters of the basin. One cautionary note in developing modeling tools is that it is just as important and challenging to develop tools to analyze model output as it is to develop a model to represent the basin. If adequate data do not exist, which is probably inevitable, the process of developing a system model can help in identifying what data are necessary to address the problems at hand.

LESSON 5: AN IMPORTANT EARLY STEP IS TO QUALITATIVELY DEFINE THE BOUNDARIES OF AN ACCEPTABLE AGREEMENT. THERE ARE MANY TYPES OF "BOUNDARIES" TO CONSIDER INCLUDING: TECHNICAL BOUNDARIES, POLITICAL BOUNDARIES, LEGAL BOUNDARIES, TRUST BOUNDARIES AND EFFICIENCY BOUNDARIES. ONCE POLICY DECISION-MAKERS HAVE QUALITATIVELY DEFINED THE BOUNDARIES OF AN ACCEPTABLE AGREEMENT, TECHNICAL STAFF CAN DEVELOP A RANGE OF RESPONSES THAT LIE WITHIN THESE BOUNDARIES

As the previous lesson pointed out, it is important to develop tools and collect data to allow technical staff to examine the water resource problems of a basin. However, just as important is to define what would be an acceptable agreement. Although the need to define an acceptable agreement seems obvious, no agreement was ever reached on what would constitute an acceptable agreement, despite over 10 years of study, data collection, tool development, and negotiating over the Allocation Formula the three states and the federal government never reached agreement on what would constitute and acceptable agreement (Leitman, 2005).

There are multiple types of boundaries between acceptable and unacceptable which need to be considered including technical boundaries, political boundaries, legal boundaries, trust boundaries, and efficiency boundaries. Technical boundaries simply refer to what actions are technically possible or feasible. This would consider issues such as the level of augmentation possible from a reservoir system for water supply or waste water dilution purposes. Political boundaries refer to what is acceptable in a political context. It is possible for a response to be acceptable in a technical context, but not acceptable in a political context. Legal boundaries simply refer to what actions can be done within the current legal framework of the negotiating parties.

Trust boundaries refer to what actions can be taken, whether legal or not legal, that would build trust among the negotiating parties. In the case of the ACF negotiations, the development of an MOU by the three governors was clearly legal, but in taking this action it broke down the trust among key stakeholders and ultimately contributed to the end of the negotiations. Efficiency refers to the timing and cost of reaching an agreement. There are limits to both and these must be accounted for when working out an acceptable agreement.

Defining the boundaries of an acceptable agreement is the responsibility of policymakers involved in the decision-making process.

Once these boundaries are defined, developing the suite of acceptable responses based on these boundaries is the responsibility of technical staff. It is important to avoid a situation where either a policymaker is making technical decisions or technical staff is making a policy decision. Either situation will most likely result in decisions that either do not work or cannot be implemented. In the ACF negotiations the failure of the decision-makers to define the boundaries of an acceptable decision forced technical staff into the dilemma of having to evaluate alternative scenarios without any guidance of what was acceptable.

4. Conclusions

All of the conclusions presented are not intended to be specific either to the ACF basin or the United States. They are general broad perceptions that may help in such disputes in Central Asia. However, ultimately the decisions of how to approach transboundary needs to be made in the basin and supported by the political power structure in the basin. Perhaps one of the major lessons to be gleaned from the ACF experience for regions such as Central Asia is that the process of negotiating and implementing a transboundary water management infrastructure is difficult and not to be taken for granted. There is one interesting parallel between the ACF basin and Central Asian region that should be considered: the relative inexperience of both regions with utilizing such structures.

The ACF Compact was the first ever in that region of the United States and the first in the nation since passage of the major environmental laws in the early 1970s. Consequently, none of the staff working on the issue had real-world experience working on such issues and consequently the effort was a "prototype" effort, instead of an experienced team working on a difficult issue. In the end the combination of inexperience and complexity led to a failure to reach an agreement on how to address the problems at hand.

In Central Asia many of the individuals who may work on such issues are probably similarly inexperienced both as to their level of expertise with negotiations and in their experience in dealing with transboundary water problems. It is therefore recommended that those who will participate in future Central Asian forums be provided the opportunity to receive intensive training both in negotiation fundamentals and in skills necessary to manage transboundary waters before they have to use these skill sets for real at the negotiation table. Efforts spent to enhance both of these skills before entering into serious negotiations will increase the chances of developing an approach to address these problems in this region.

References

Dellapenna, J.W., 2006. The Law, Interstate Compact, and the Southeastern Water Compact. Interstate Water Allocation in Alabama, Florida and Georgia: New Issues, New Methods, New Models, 51–77. University Press of Florida.

Dellapenna, J.W., J.L. Jordan, S. Leitman and A.T. Wolf., 2006. Conclusions, Outcomes, Updates and Lessons Learned. Interstate Water Allocation in Alabama, Florida and Georgia: New Issues, New Methods, New Models, 233–248. University Press of Florida.

Kenney, D.S., 1997. Review of Coordination Mechanisms with Water Allocation Responsibilities. Paper prepared for the ACT–ACF Comprehensive Study. Mobile, AL. US Army Corps of Engineers, Mobile District.

Leitman, S. 2005, Negotiations of a Water Allocation Formula for the Apalachicola–Chattahoochee–Flint Basin. Adaptive Governance, 74–88. Resources for the Future.

Palmer R. and others., 1999. Modeling Water Resources Opportunities, Challenges and Tradeoffs: The Use of Shared Vision Modeling for Negotiation and Conflict Resolution. Proceedings of 1999 ASCE Conference.

Stephenson, K., 2002. The what and why of Shared Vision Planning for Water Supply. Speech prepared for the panel session "Collaborative Water Supply Planning: A Shared Vision Approach for the Rappahannock River Basin" Water Security in the 21st Century Conference, Washington, DC. July 30, 2002.

US Army Corps of Engineers. 1989. Post Authorization Change Notification Report for the Re-allocation of Storage from Hydropower to Water Supply at Lake Lanier, Georgia. Mobile District, US Army Corps of Engineers.

Wolf, Aaron 2001, Transboundary Waters: Shared Benefits, Lessons Learned. Report to the Secretariat of the International Conference on Freshwater.

DETERMINING EQUITABLE UTILIZATION OF TRANSBOUNDARY WATER RESOURCES: LESSONS FROM THE UNITED STATES SUPREME COURT[1]

GEORGE WILLIAM SHERK[2]
Colorado School of Mines
Stratton Hall, Room 309
Golden, CO 80401-1887 USA

Abstract: This paper is a brief discussion on the equitable utilization of transnational water using lessons learned from the United States Supreme Court rulings. Development and adoption of the UN. Convention on the Law of the Non-Navigational Uses of International Watercourses (the Watercourses Convention) is discussed in Section 3. The principle of equitable utilization contained in the Watercourses Convention, particularly as that principle relates to the equitable apportionment decisions of the US Supreme Court, is reviewed in Section 4. Conclusions are presented in Section 5.

Keywords: legal issues, conflict resolution, transnational watercourses, equitable utilization principle

---

[1] Paper presented at the NATO Advanced Research Workshop on Facilitating Regional Security in Central Asia through Improved Management of Transboundary Water Basin Resources, Almaty, Kazakhstan, 20–22 June 2006.

[2] D.Sc., School of Engineering and Applied Science, The George Washington University; J.D., University of Denver College of Law; M.A., B.A., Colorado State University. Dr. Sherk is the Hennebach Program in the Humanities Visiting Scholar at the Colorado School of Mines, Golden, Colorado. He is also an Adjunct Professor at the University of Denver College of Law, Denver, Colorado and an Associate at the International Water Law Research Institute, University of Dundee, Dundee, Scotland. He may be reached at gsherk@mines.edu or gwsherk@h2olaw.com.

1. Introduction

More than 300 river basins covering in excess of 50% of the total land area of the Earth transcend national boundaries. For example, the Danube River is shared by seventeen countries while the Nile River is shared by nine countries.

As a result, particularly in times of scarcity, there is significant potential for conflict over transboundary water resources. This potential gives rise to a number of questions: Is the upstream state entitled to use of all the water that originates on its territory? Are the prior developments of downstream states protected against subsequent uses of their upstream neighbors? How can such conflicts of uses be resolved?

The following section contains a brief discussion of the potential for transnational conflicts over water. Development and adoption of the UN Convention on the Law of the Non-Navigational Uses of International Watercourses (the Watercourses Convention) is discussed in Section 3. The principle of equitable utilization contained in the Watercourses Convention, particularly as that principle relates to the equitable apportionment decisions of the US Supreme Court, is reviewed in Section 4. Conclusions are presented in Section 5.

2. The Potential for Transnational Conflicts Over Water

Numerous states have based claims to transnational water resources on theories of absolute territorial sovereignty, absolute territorial integrity or limited territorial sovereignty. Assertion of claims to transboundary water resources based on these theories presents a significant potential for transnational conflict regarding shared water resources. In order to address such conflicts (and prevent them, hopefully), it is essential to understand the relative rights and duties of riparian states with regard to their utilization of transboundary water resources.

3. The UN Convention on the Law of the Non-Navigational Uses of International Watercourses[3]

In 1970, the General Assembly of the United Nations recommended that the International Law Commission of the United Nations (the ILC) "take up the study of the law of the non-navigational uses of international watercourses with a view to its progressive development and codification". After nearly a quarter century of study and deliberation, the ILC adopted a set of draft articles on the non-navigational uses of transnational watercourses. The draft articles adopted by the ILC were referred to the UN General Assembly to be used as a starting point for the drafting of a multilateral water convention.

In 1996, the UN's 6th Committee, convened as the Working Group of the Whole and commenced meetings on the draft articles. The first two weeks of meetings revealed the extent of controversy that existed on key issues. At the end of this first session in November of 1996, there was substantial uncertainty as to whether agreement on a text was possible.

Following much debate, many proposals and the inevitable compromises, the Working Group of the Whole at the second two-week session in March/April of 1997 took the unusual step of voting on a revised draft text. By a vote of 42 states for and three states against (with 18 state abstentions), a final text was adopted by the Working Group of the Whole.

The issues central to the controversy in the Working Group arose in three key areas: (i) To what extent did states have to comply with the provisions of the Convention regarding existing and future watercourses agreements, (ii) What was to be the substantive content and relationship between the principles of equitable utilization (Article 5) and no significant harm (Article 7) and (iii) to what extent were states to be bound by dispute settlement mechanisms? The compromise reached in each of these areas reveals a central ground acceptable to the majority of states.

---

[3] This section is adapted and condensed from Sherk, Wouters and Rochford, "Water Wars in the Near Future? Reconciling Competing Claims for the World's Diminishing Freshwater Resources: The Challenge for the Next Millennium" in *Managing the Era of Great Change.* Alexandria, VA: International Strategic Studies Association (1998). Reprinted at 3 *The CEPMLP On-Line Journal*, http://www.dundee.ac.uk/cepmlp/journal/html/vol3/article3-2.html.

On the first issue, the final text affords states substantial flexibility with respect to existing and future watercourse agreements. States are free to "adjust the provisions" of the Convention to the particular characteristics of the watercourse involved, so long as the rights of other watercourse states are not affected by the Convention. With respect to dispute settlement, once again states are afforded ample latitude, although the revised text called for compulsory fact-finding which, upon scrutiny, reveals a procedure closer to a compulsory conciliation procedure. On the crucial issue most relevant to this paper, the Working Group made substantial revisions to the formulation of the no significant harm rule contained in Article 7 of the draft with the result being that the principle of equitable utilization emerged as the governing rule of the Convention.[4]

The final text adopted by the Working Group of the Whole was appended to a draft resolution put forth before the UN General Assembly by 33 states on 21 May 1997. On 23 May 1997 the UN General Assembly adopted the Convention on the Law of the Non-Navigational Uses of International Watercourses. Containing 37 articles with a 14-article Annex, the instrument was adopted by a vote of 104 states in favor and three states against (with 26 state abstentions). The text was opened for signature from that date until 20 May 2000.

The adoption of this framework convention, including the process with which this was achieved raises important issues relevant to the future management of transnational water resources. The Watercourses Convention provides important substantive and procedural rules for states to follow regarding such water resources. The overall aim of the instrument is to provide realistic means to prevent or resolve disputes over water. Despite controversy on some key issues, states have supported the adoption of this body of rules at two critical stages in the development of the Watercourses Convention: First, by

---

[4] The no significant harm rule, significantly revised from its former versions contained in the 1991 and 1994 ILC draft articles, can be read as subsidiary to the equitable utilization principle contained in Article 5. Article 7(2) provides: "Where significant harm nevertheless is caused to another watercourse state, the states whose use causes the harm shall, in the absence of agreement to such use, take all appropriate measures, having due regard for the provisions of articles 5 and 6, in consultation with the affected state, to eliminate or mitigate such harm and, where appropriate, to discuss the question of compensation." These provisions replaced the 1994 ILC draft Article 7 which read: "States shall exercise due diligence to utilize an international watercourse in such a way as not to cause significant harm to other watercourse States."

the majority of states voting in the Working Group of the Whole. Second, by the majority of states voting at the UN General Assembly.

As with the approach adopted in the Watercourses Convention, under the ILA's approach, "[w]hat is a reasonable and equitable share ... is to be determined in the light of all the relevant factors in each particular case." It is in this context, a determination of "all relevant factors," that the equitable apportionment decisions of the US Supreme Court have great relevance.[5]

4. Determining Equitable Utilization: Lessons from the United States Supreme Court

The United States Constitution, article 3, section 2, provides that "In all cases ... in which a state shall be a Party, the Supreme Court shall have original jurisdiction." In litigation between states, the Supreme Court's jurisdiction is both original and exclusive.[6]

The first interstate water conflict addressed by the Supreme Court was the conflict between the States of Kansas and Colorado over the waters of the Arkansas River. In 1902, the Court concluded that it had jurisdiction over the conflict and that the doctrine of *parens patriae* allowed each state to represent its citizens in litigation before the Supreme Court.[7] The basis for this, the Court reasoned, was that the states "cannot make war upon each other."[8]

It is clear that the Supreme Court in 1902 saw itself as applying principles of international law: "Sitting, as it were, as an international,

---

[5] As McCaffrey has noted, the principle of equitable utilization was "[b]orn of the US supreme Court's decisions in interstate apportionment cases beginning in the early 20th century, and supported by decisions in other federal states[.]" McCaffrey, *The Law of International Watercourses* 322. Oxford: Oxford University Press (2001) (internal citations omitted).

[6] 28 U.S.C. § 1251(a) provides that "[t]he Supreme Court shall have original and exclusive jurisdiction of all controversies between two or more States." See generally Note, "The Original Jurisdiction of the United States Supreme Court," 11 *Stanford Law Review* 665, 681–683 (1959).

[7] *Kansas v. Colorado*, 185 US 125 (1902). With regard to conflicts between the states, the essence of the *parens patriae* doctrine is that a state speaks for all of that state's citizens. The doctrine is based on the common law doctrine that the sovereign held a guardianship over both the nation and the citizens of the nation. "At common law the king is parens patriae, father of his country, which is but the medieval mode of putting what we mean today when we say that the state is the guardian of social interests." Pound, *The Spirit of the Common Law* (1921) cited in Sherk, "Interstate Water Conflicts and Individual Rights: Perspectives from the United States Supreme Court," 25 *Water International* 519 (2000).

[8] 185 US at 143.

as well as a domestic tribunal, we apply Federal law, state law, and international law, as the exigencies of the particular case may demand[.]"[9] Now, over a century later, the decisions of the Supreme Court have themselves become a source of international law.

Following the 1902 decision in *Kansas v. Colorado*, the Court has exercised its original jurisdiction on numerous occasions to resolve interstate water conflicts.[10] These decisions provide the basis for the doctrine of equitable apportionment. This is the doctrine that would be applied by the Court to conflicts between states regarding shared water resources should those conflicts be brought to the Court for resolution.[11] It is also the doctrine that is reflected in the substantive requirements of the concept of equitable utilization as embodied in the 1997 Watercourses Convention.

Through its equitable apportionment decisions, the Supreme Court has identified eleven interrelated factors that should be considered in the resolution of an interstate water conflict. The Supreme Court has also identified five factors that should not be considered in the resolution of such conflicts. It must be noted, however, that not all of these factors are relevant to every interstate water conflict.

Factors that should be considered in the resolution of an interstate water conflict:

A. WATER CONSERVATION AND THE AVOIDANCE OF WASTE

A review of the Court's equitable apportionment decisions makes it clear that water conservation and the avoidance of waste is one of the most important factors to be considered when addressing interstate water conflicts. The duty of each state to "exercise her right

---

[9] 185 US at 146–147.

[10] These decisions are summarized in Sherk, *Dividing the Waters: The Resolution of Interstate Water Conflicts in the United States.* London: Kluwer Law International (2000). It should be noted that the Supreme Court has also addressed a number of interstate water conflicts in the context of disputes over the requirements of interstate water compacts. While not equitable apportionment actions *per se*, many of the requirements of the doctrine have been applied by the Court to the interstate compact cases.

[11] *See generally* Sherk, "Equitable Apportionment After *Vermejo*: The Demise of a Doctrine," 29 *Natural Resources Journal* 565 (1989); Tarlock, "The Law of Equitable Apportionment Revised, Updated, and Restated," 56 *University of Colorado Law Review*. 381 (1985); Note, "*Colorado v. New Mexico II*: Judicial Restraint in the Equitable Apportionment of Interstate Waters, 62 *University of Denver Law Review* 857 (1985); Note, "Is There a Future for Proposed Water Users in Equitable Apportionment Suits?" 25 *Natural Resources Journal* 791 (1985).

reasonably and in a manner calculated to conserve the common supply" of water was established in *Wyoming v. Colorado*.[12]

The existence of such a duty was reaffirmed in *Colorado v. New Mexico*.[13] In a decision involving the waters of the Vermejo River, the Court ruled that states have "an *affirmative* duty to take reasonable steps to conserve and augment the water supply of an interstate stream."[14]

In essence, the Court's decisions make it clear that wasteful or inefficient uses would not be protected. Furthermore, the requirements that waste be prevented and that water be used efficiently are applicable to both existing and new facilities.

B. POTENTIAL EFFECT OF WASTEFUL USES ON DOWNSTREAM AREAS

Consideration of the potential effects of wasteful uses on downstream areas was one of the factors enumerated by the Court in *Nebraska v. Wyoming*.[15] This factor reflects the realities of the western United States in that one person's waste is another person's water supply. It also reflects a fundamental tenet of the prior appropriations water law doctrine that an appropriator is entitled to the maintenance of stream conditions substantially as they were at the time the appropriation was initiated. If, for example, stream flows being utilized in a downstream area included flows resulting from wasteful uses in an upstream area, it may not be equitable to require an upstream area to implement water conservation measures and then allow the upstream area to claim the use of the quantity of water conserved.

C. AUGMENTATION OF WATER SUPPLY

As noted above, states have an affirmative duty to augment the water supplies of interstate streams. Such augmentation could include water recycling or water reuse programs. In appropriate circumstances, it might also include weather modification.[16]

---

[12] 259 US 419, 484 (1922).

[13] 459 US 176 (1982).

[14] 459 US at 185 (emphasis added).

[15] 325 US 589, 618 (1945).

[16] For example, the seeding of orographic clouds during the winter in Colorado has been proposed as one means of augmenting existing water supplies. See generally Danielson, Sherk and Grant, "Legal System Requirements to Control and Facilitate Water Augmentation

D. EXTENT OF ESTABLISHED USES

This was another of the factors enumerated by the Court in *Nebraska v. Wyoming*.[17] It should not be assumed, however, that an equitable apportionment of interstate water resources will protect established uses in all instances.

Protection of established uses must be considered together with the requirements of water conservation and the efficient use of water. As the Court noted in *Colorado v. New Mexico*, protection of an existing economy "will usually be compelling" but not always.[18] Wasteful or inefficient established uses will not be protected.

E. CONSUMPTIVE USES OF WATER IN DIFFERENT STREAM SEGMENTS

Another of the factors enumerated in *Nebraska v. Wyoming* was the need to consider the consumptive use of water in different stream segments.[19] Consideration of this factor requires a number of questions to be asked and answered, specifically: Who is using what quantity of water? Where is the water being used? When is it being used? Is it being used efficiently?

These are threshold questions that must be addressed whenever conflicts over shared water resources are being addressed. If these questions have not been asked and answered, then there is no factual foundation upon which conflict management may be constructed.

F. CHARACTER OF RETURN FLOWS

In the United States, the primary source of nitrogen and phosphorus contaminating surface waters is return flows from irrigated agriculture. Consideration of the character of such return flows was another of the factors enumerated by the Court in *Nebraska v. Wyoming*.[20]

For example, if irrigation return flows in an upper basin state were high in nitrogen and phosphorus, then equity might require higher

---

in the Western United States," 6 *Denver Journal of International Law and Policy* 511 (1976), reprinted at *Water Needs for the Future* 289. Boulder, CO: Westview Publishing Co. (V. Nanda, ed., 1977).

[17] 325 US at 618.

[18] 459 US 176, 187 (1982) (emphasis added).

[19] 325 US at 618.

[20] 325 US at 618.

downstream flows in order to provide the capacity needed to assimilate the wastes.

G. RATE OF RETURN FLOWS

Directly related to the character of return flows, the rate of return flows addresses the questions noted above: What quantity of water is available, where is it available and when is it available? [21]

H. AVAILABILITY OF STORAGE WATERS

One of the factors to be considered in the equitable apportionment of interstate water resources is the relative capacity of the states to conserve the shared water resource. This was another of the factors enumerated by the Court in *Nebraska v. Wyoming*.[22] A state having the capacity to conserve the common supply of water may have an obligation to do so.

This is especially relevant in terms of water supply shortages or interruptions. States having the ability to store water are much more capable of responding to water supply shortages or interruptions than are states lacking the capacity to store water. It may not be equitable, for example, to require states to share water shortages on a *pro rata* basis if one state has the capability to store water and another state does not.

I. PHYSICAL CONDITIONS

Hydrogeologic conditions existing in the watershed much also be considered. For example, in *Washington v. Oregon*,[23] the Court addressed a conflict involving use of the water in a tributary of the Walla Walla River. Washington, the downstream state, sought to enjoin upstream diversions in Oregon. The Court allowed the diversions in Oregon to continue because the hydrogeologic conditions were such that the water would not have reached Washington even if the diversions had been eliminated.[24]

---

[21] The rate of return flow was another of the factors enumerated by the Court in *Nebraska v. Wyoming*, 325 US at 618.

[22] 325 US at 618.

[23] 297 US 517 (1936).

[24] Consideration of physical conditions was another of the factors enumerated by the Court in *Nebraska v. Wyoming*, 325 US at 618.

J. CLIMATIC CONDITIONS

As noted above, one of the factors to be considered is the availability of storage waters. This is directly related to a state's capability to respond to changing climatic conditions.[25]

For example, in *Wyoming v. Colorado*,[26] the Court addressed the question of which state was better suited to bear the burden of climate variability. Concluding that Wyoming (the downstream state) was better suited, the Court ruled that Wyoming had to bear the burden of shortages in low water years but was entitled to surplus flows during high water years.

K. DAMAGES TO RESPECTIVE STATES IF LIMITATIONS ARE IMPOSED

This was the final factor enumerated by the Court in *Nebraska v. Wyoming*.[27] In essence, what are the impacts of the decision? In order to equitably apportion a shared water resource, the Court must "balance the equities" including a balancing of benefits and costs.

Factors that should not be considered in the resolution of an interstate water conflict:

(a) Requirements of State Law

As noted above, the first interstate water conflict heard by the Court was the conflict between Kansas and Colorado over the waters of the Arkansas River. One of the issues before the Court was a conflict between the water law systems adopted by the two states. Kansas, a riparian doctrine state, argued that the riparian water law doctrine should apply. Colorado, a prior appropriation doctrine state, argued that the prior appropriation doctrine should apply.

The Court answered the question by ruling that it would not be bound by the requirements of state law. Specifically, "[t]he determination of the relative rights of contending states in respect of the use of streams flowing through them does not depend upon the same considerations and is not governed by the same rules of law that are applied in such states for the solution of similar questions of

---

[25] Climatic conditions was another of the factors enumerated by the Court in *Nebraska v. Wyoming*, 325 US at 618.

[26] 259 US 419 (1922).

[27] 325 US at 618.

private right."[28] In a later ruling in the same case, the Court concluded that principles of equity control when state water laws are in conflict.[29]

(b) Area of Origin Entitlement

It has been argued frequently that a headwaters state should be entitled to a portion of an interstate stream simply because the stream arose in that state.[30] This argument was addressed and rejected by the Court in the conflict between Colorado and New Mexico over the waters of the Vermejo River.

In its first Vermejo decision, the Court concluded: "If the Special Master believed that Colorado was entitled to use of the Vermejo River simply because the river arose in Colorado, the Special Master erred."[31] Any such area of origin preference, the Court ruled, was "inconsistent with our emphasis on flexibility in equitable apportionment."[32]

This conclusion was reaffirmed two years later in the second Vermejo decision: "[T]he equitable apportionment of appropriated rights should turn on the benefits, harms, and efficiencies of competing uses, and the source of the Vermejo River's water should be essentially irrelevant to the adjudication of these sovereigns' competing claims."[33]

(c) Downstream Entitlement to Undiminished or "Natural" Streamflows

As upper basin states are not entitled to an area of origin preference, downstream states are not entitled to undiminished or "natural" stream flows. This contention, which arose primarily in the context of the

---

[28] 185 US 125, 146 (1902):

[29] 206 US 46 (1907). With regard to the requirements of state water laws, the Court did rule in *Nebraska v. Wyoming*, that prior appropriation is a "basic" or "guiding" principle of equitable apportionment. 325 US at 618.

[30] For example, the consistent refusal of the State of Georgia to acknowledge that it does not have a priority right to the waters of four streams arising in Georgia but shared with the States of Alabama and Florida has been one of the causes of an ongoing interstate water conflict. See Sherk, "The Management of Interstate Water Conflicts in the 21st Century: Is it Time to Call Uncle?" 12 *New York University Environmental Law Journal* 764, 811–812 (2005).

[31] *Colorado v. New Mexico*, 459 US 176, 181 n. 8 (1982).

[32] *Ibid.*

[33] *Colorado v. New Mexico*, 467 US 310, 323 (1984).

riparian water law doctrine, was rejected by the Court in New Jersey v. New York.[34]

(d) Area of Origin Limitations

Justice William O. Douglas, in his dissent in Arizona v. California,[35] argued that the Court should consider the extent to which transbasin diversions were allowed.[36] Justice Douglas based his argument on the fact that transbasin diversions remove shared water resources from their basin of origin, often to the detriment of downstream states.

However, when confronted with this issue in Wyoming v. Colorado,[37] the Court concluded that transbasin diversions should be allowed since they were allowed under the laws of both states involved in the conflict. This conclusion was reaffirmed in New Jersey v. New York,[38] a decision in which the Court concluded that diversions need not be restricted to the basin of origin.

(e) Percentage of Shared Watershed

Also in his Arizona v. California dissent, Justice Douglas argued that the Court should consider the percentage of shared water resources falling with the specific states.[39] The Court did not (and has not) accepted Justice Douglas' argument. It is quite likely that the Court's refusal to accept this argument is based on the aforementioned need for flexibility in equitable apportionment actions.

5. Conclusions

The potential for transnational conflict over water is great. One of the essential mechanisms necessary to prevent "water wars" is the establishment of clear "rules of the game". The Watercourses Convention goes a long way in achieving this purpose. The governing principle of equitable utilization levels the playing field and offers every state an opportunity to develop transnational water resources. As

---

[34] 283 US 336 (1931).
[35] 373 US 546 (1963).
[36] 373 US at 627.
[37] 259 US 419 (1922).
[38] 283 US 336 (1931).
[39] 373 US at 627.

noted above, "all relevant factors" must be weighed in any determination of equitable utilization.

Clearly, the preferred resolution is one arrived at by agreement. Where each side knows that its concerns must be considered in the context of overall impacts in a transnational context, compromises will be easier to achieve. A number of recent international agreements relating to transboundary waters endorse the approach adopted in the Watercourses Convention. For those states that voted against the Convention, or that are not party to watercourse agreements, the weight of the growing consensus of the international community will carry persuasive force. It is now left to the international community to endorse the principles outlined in the Watercourses Convention. This would be consistent with significant state practice already in existence and would also contribute to the peaceful management of transnational water resources.

References

Danielson, Sherk and Grant, "Legal System Requirements to Control and Facilitate Water Augmentation in the Western United States," 6 Denver Journal of International Law and Policy 511 (1976), reprinted at Water Needs for the Future 289. Boulder, CO: Westview Publishing Co. (V. Nanda, ed., 1977).

McCaffrey, The Law of International Watercourses. Oxford: Oxford University Press (2001).

Note, "Colorado v. New Mexico II: Judicial Restraint in the Equitable Apportionment of Interstate Waters, 62 University of Denver Law Review 857 (1985).

Note, "Is There a Future for Proposed Water Users in Equitable Apportionment Suits?" 25 Natural Resources Journal 791 (1985).

Note, "The Original Jurisdiction of the United States Supreme Court," 11 Stanford Law Review 665 (1959).

Sherk, Dividing the Waters: The Resolution of Interstate Water Conflicts in the United States. London: Kluwer Law International (2000).

Sherk, "Equitable Apportionment after Vermejo: The Demise of a Doctrine," 29 Natural Resources Journal 565 (1989).

Sherk, "Interstate Water Conflicts and Individual Rights: Perspectives from the United States Supreme Court," 25 Water International 519 (2000).

Sherk, "The Management of Interstate Water Conflicts in the 21st Century: Is it Time to Call Uncle?" 12 New York University Environmental Law Journal 764 (2005).

Sherk, Wouters and Rochford, "Water Wars in the Near Future? Reconciling Competing Claims for the World's Diminishing Freshwater Resources: The Challenge for the Next Millennium" in Managing the Era of Great Change.

Alexandria, VA: International Strategic Studies Association (1998). Reprinted at 3 The CEPMLP On-Line Journal, http://www.dundee.ac.uk/cepmlp/journal/html/vol3/article3-2.html.

Tarlock, "The Law of Equitable Apportionment Revised, Updated, and Restated," 56 Unssiversity of Colorado Law Review. 381 (1985).

IMPROVING TRANSBOUNDARY RIVER BASIN MANAGEMENT BY INTEGRATING ENVIRONMENTAL FLOW CONSIDERATIONS

KARIN M. KRCHNAK
Senior International Water Policy Advisor
The Nature Conservancy
4245 North Fairfax Drive
Arlington, VA 22203 USA

Abstract: Freshwater ecosystems provide a wealth of services to humans including food and fiber, water purification, fish and wildlife habitat, tourism and recreational opportunities, shipping routes, employment, and opportunities for cultural and spiritual renewal. To provide this range of services freshwater systems depend on the cycling of water and on functioning ecological processes and species assemblages. Water management has traditionally focused on meeting the needs and desires of a growing and changing human population without due consideration to the needs and limits of our freshwater systems. It is suggested that the first priority in any freshwater allocation scheme should be to make an "ecosystem support allocation. The ESWM framework understands the flow necessary to sustain or restore the integrity of a river, assesses human influence on water flow, and identifies areas of incompatibility between people and nature to provide a foundation for next steps.

Keywords: freshwater ecosystems, hydrologic cycle, environmental flows, water allocation, sustainability, environmentally sustainable water management

1. Introduction

The hydrologic cycle on Earth supports an abundance of life, including human life. Freshwater ecosystems are supported by the hydrologic cycle, providing immeasurable services and benefits to humans, ranging from food and water purification to spiritual renewal. Despite growing awareness of the importance of healthy freshwater ecosystems, human actions continue to degrade the freshwater ecosystems upon which we depend. Even with the policy movement toward integrated water resource management, the integration of ecosystem considerations in water management remains largely neglected.

The Millennium Ecosystem Assessment notes that: "Any progress achieved in addressing the MDGs[1] of poverty and hunger eradication, human health, and environmental protection is unlikely to be sustained if most of the ecosystem services on which humanity relies continue to be degraded."[2] The Assessment also clearly points out "The use of…freshwater…is now well beyond levels that can be sustained even at current demands, much less future ones."[3] Reasons for this state regarding water resources involves a number of factors, including: the failure to place economic value on freshwater services, the desire by governments and donor institutions to focus on problems that have a perceived silver bullet solution, inadequate funding, the relative political weakness of ministries of environment and water compared to ministries that finance infrastructure development, and a misperception that water allocated to the environment is water unavailable for humans. In reality, and despite these challenges, to achieve long-term sustainability and improve human well-being worldwide, we must value and conserve freshwater ecosystems.

This paper begins with brief descriptions of the hydrological cycle and the importance of freshwater ecosystems and the environmental services they provide. It then outlines the range of impacts on freshwater ecosystems that result from mismanagement of water. The paper then describes global policy development for an ecosystem-based approach to decision-making. The remaining part of the paper describes tools for addressing environmental flows. It specifically describes The Nature Conservancy's approach in influencing hydropower operations by integrating environmental flow considerations.

---

[1] Millennium Development Goals. See http://www.un.org/millenniumgoals/
[2] Millennium Ecosystem Assessment, **Ecosystems and Human Well-being: Synthesis**, Island Press (2005).
[3] Millennium Ecosystem Assessment, **Ecosystems and Human Well-being: Synthesis**, Island Press (2005).

2. The Importance of Freshwater Ecosystems

Ecosystem services are defined as a variety of culturally and socially-valued goods and services that human society derives from natural ecosystems.[4] Freshwater ecosystems[5] provide a wealth of food and fiber, water purification, fish and wildlife habitat, tourism and recreational opportunities, shipping routes, employment, and opportunities for cultural and spiritual renewal. People across the globe depend on fishes as their primary source of protein, with some regions particularly reliant on fish due to the fundamental social and economic role of fisheries.[6] Moreover, the genetic and chemical components of aquatic species may offer humans invaluable pharmaceutical and other benefits.[7] However, to provide the range of services on which humans depend, freshwater systems themselves depend on the cycling of water and on functioning ecological processes and species assemblages.

The hydrological cycle – the Earth's method of recycling our water supply – acts like a giant water pump that continually transfers freshwater from the oceans to the land and back again. In this solar-driven cycle, water evaporates from the Earth's surface into the atmosphere and is returned to the Earth as rain or snow. Part of this precipitation evaporates back into the atmosphere, while another part flows into streams, aquifers, rivers and lakes, commencing a journey back toward the sea. Still another part sinks into the soil and becomes

---

[4] For more information on ecosystem services, see Millennium Ecosystem Assessment, **Ecosystems and Human Well-being: Synthesis**, Island Press (2005). See also the United Nations', **World Water Development Report**, Chapter 6 (2003). Malin Falkenmark describes ecosystems as: "a set of interacting organisms and the solar driven system that they compose, comprising both primary producers, and consumers and decomposers. In combination they mediate the flow of energy, the cycling of elements (including water) and spatial and temporal patterns of vegetation. An ecosystem may be of any scale from global all the way down to local." Falkenmark, M., **Water Management and Ecosystems: Living with Change**, TEC Paper #9, GWP (2003).

[5] Freshwater ecosystems are some of the most complex. They include rivers, lakes, streams, marshes, swamps, other wetlands, coastal bays, estuaries, aquifers, and deltas.

[6] www.fao.org

[7] Omernick, J.M. and Bailey, R.G. "Distinguishing between watersheds and ecoregions," Journal of the American Water Resources Association, 33: 935–949 (1997).

soil moisture or gets stored as groundwater.[8] Much of the world's groundwater slowly works its way back into the flow of surface water.[9]

This water pump can operate in sporadic and unpredictable ways. The available water supply is not distributed evenly around the Earth, throughout the seasons, or from year to year. Often, sufficient quantities of water are not available where and when humans need them. In other cases, we have too much water in the wrong place at the wrong time. The hydrological cycle does not always cooperate with human's needs and desires for water.[10]

Our freshwater ecosystems require certain quality and quantity of water to maintain their health, including the thousands of species and activities they support. Approximately 40% of all fish are freshwater species,[11] with some 200 new freshwater species being identified each year.[12] However, at least 20% of all known freshwater fish – some 2,000 species out of the 10,000 so far identified – are endangered, vulnerable or extinct.[13]

Environmental flows are the amounts of freshwater flowing in rivers and streams and into bays and estuaries that are needed to support healthy, diverse populations of fish and other aquatic life. Variable flow conditions are required to enable aquatic species to

---

[8] Hinrichsen, D., et. al., **Population, Water & Wildlife: Finding a Balance**, National Wildlife Federation (2003).

[9] "After reaching the land surface the rainwater is partitioned into the green water vapour flow supporting the terrestrial ecosystems and the blue water liquid flow supporting the aquatic ecosystems and accessible for human use." Falkenmark, M., **Water Management and Ecosystems: Living with Change**, TEC Paper #9, GWP (2003).

[10] Hinrichsen, D., et. al., **Population, Water & Wildlife: Finding a Balance**, National Wildlife Federation (2003).

[11] Abramovitz, J.N., "Imperiled Waters, Impoverished Future: The Decline of Freshwater Ecosystems," **World Watch Paper** 128 (1996).

[12] Revenga, C., and Y. Kura, **Status and Trends of Biodiversity of Inland Water Ecosystems**, CBD Technical Series No. 11, CBD Secretariat, Montreal, Canada (2003).

[13] Moyle, P.B. and R.A. Leidy, "Loss of biodiversity in aquatic ecosystems: Evidence from fish faunas," **Conservation Biology: The Theory and Practice of Nature Conservation, Preservation and Management**, Chapman and Hall, NY (1992).

complete their life cycles.[14] For example, fish may depend on flooding conditions for upstream migrations for reproduction and stable low flows to conserve their energy during cold winters. Different flow conditions also enable fish species to access critical areas for feeding and shelter.[15] Quantity and quality issues are interlinked as sufficient flows not only sustain the social, ecological, and hydrological functions of watersheds and wetlands but also offset pollution.[16] In addition, groundwater and surface waters are interlinked; thus, degradation or diminution of groundwater may affect surface waters and the associated ecosystems. Unfortunately, despite improvements in information, insufficient data exist on water quantity and quality requirements of freshwater ecosystems in many river basins.

3. Impacts on Freshwater Ecosystems from Mismanagement of Water Resources

Water management has traditionally focused on meeting the needs and desires of a growing and changing human population without due consideration to the needs and limits of our freshwater systems.[17] As a result, freshwater and related ecosystems have felt the impacts of mismanagement, with some rivers such as the Rio Grande in the USA and Mexico now failing to reach the sea in many years or during droughts. Globally, it is estimated that the world has already lost half of its wetlands with most of the destruction having taken place in the

---

[14] Richter, BD, et. al., "How much water does a river need?" **Freshwater Biology**, 37:231–249 (1997); Poff, NL, et. al., "The natural flow regime: a paradigm for river conservation and restoration," **BioScience**, 47:769–784 (1997); Richter, BD, et. al, "Ecologically sustainable water management: managing river flows for ecological integrity," **Ecological Applications,** 13:206–224 (2003).

[15] Falkenmark, M., **Water Management and Ecosystems: Living with Change**, TEC Paper #9, GWP (2003).

[16] Davis, R., and Hirji, R. (Ed.), **Water Quality: Assessment and Protection**, Water Resources and Environment, Technical Note D.1., World Bank (2003). Naturally vegetated riparian zones help trap sediments and break down nonpoint source pollution. Master, Lawrence, et. al. (Ed.), **Rivers of Life: Critical Watersheds for Protecting Freshwater Biodiversity**, The Nature Conservancy (1998). Aquatic ecosystems "cleanse" on average 80% of their global incident nitrogen loading. Millennium Ecosystem Assessment, **Ecosystems and Human Well-being: Synthesis**, Island Press (2005).

[17] "Five to possibly 25% of global fresh water use exceeds long-term accessible supplies." Millennium Ecosystem Assessment, **Conditions & Trends Assessment: Freshwater**, World Resources Institute (DRAFT 2005).

last 50 years during the period of rapid population growth and industrialization.[18] The destruction of aquatic habitat from development and urban sprawl, the construction of dams, excessive surface and groundwater withdrawals, and pervasive pollution has decimated natural freshwater systems.[19] It is not only freshwater species that are rapidly disappearing, but also saltwater and terrestrial species that depend on healthy freshwater ecosystems for their survival.[20]

The approximate 70% of global water use for agriculture often masks the other impact of agriculture – pollution of our waters. Water pollution is a particularly difficult problem in countries where population is growing rapidly, development demands are great and governments have limited resources for investment in water management and pollution control.[21] Both agricultural and industrial pollution is undermining natural systems. Pesticides like DDT, aldrin and toxaphene, and industrial chemicals like PCBs and hexachlorobenzene wipe out biodiversity and affect human health. In developing countries, on average, 90–95% of all domestic sewage and 75% of all industrial waste are discharged into surface waters without any treatment whatsoever.[22] The environment and human health are impacted as a result of this, particularly as the amounts overwhelm the assimilative capacity of waterways. Unfortunately, the costs of biodiversity losses

---

[18] Barbier, E.B. "Sustainable Use of Wetlands – Valuing Tropical Wetland Benefits: Economic Methodologies and Applications," **The Geographical Journal**, 159, 1 (1993).

[19] The Millennium Ecosystem Assessment notes that for "freshwater ecosystems and their services, depending onthe region, the most important direct drivers of change in the past 50 yrs include modification of water regimes, invasive species, and pollution, particularly high levels of nutrient loading." Millennium Ecosystem Assessment, **Ecosystems and Human Well-being: Synthesis**, Island Press (2005). This section briefly describes these but also it is important to add climate change which is not discussed in this chapter.

[20] Hinrichsen, D., et. al., **Population, Water & Wildlife: Finding a Balance**, National Wildlife Federation (2003).

[21] Hinrichsen, D., et. al., **Population, Water & Wildlife: Finding a Balance**, National Wildlife Federation (2003).

[22] Carty, W, "Towards an Urban World," **Earthwatch** 43: 2–4 (1991); Millennium Ecosystem Assessment, **Conditions & Trends Assessment: Freshwater**, World Resources Institute (DRAFT 2005).

are difficult to quantify,[23] making costly investments in waste management systems a low priority.

As more people move to cities, governmental officials seek ways to meet the water needs of the growing human population but often fail to adequately consider the long-term health of freshwater ecosystems to support the growing urban areas.[24] Currently, 47% of the world's people reside in towns and cities, a percentage that is expected to increase to 55% by 2015.[25]

4. Policy Development at the Global Level to Protect Freshwater Ecosystems

For decades, global policymakers have focused on improving water management as part of the overall sustainable development agenda. The Dublin Statement of the International Conference on Water and the Environment of 1992 is often referred to for its emphasis on a holistic approach to water management.[26] At the United Nations Conference on Environment and Development, governments delineated the way forward on sustainable development, including water conservation. The very first paragraph of Chapter 18 of Agenda 21[27] recognizes the importance of ecosystem health and the role of the hydrological cycle. Governments are called upon to "make certain that adequate supplies of water of good quality are maintained for the entire population of this planet, while preserving the hydrological, biological and chemical functions of ecosystems, adapting human activities within the capacity limits of nature...."[28] As part of a movement toward integrated water resource management, Chapter 18 also observes that "water resources have to be protected, taking into

---

[23] Davis, R., and Hirji, R. (Ed.), **Water Quality: Wastewater Treatment**, Water Resources and Environment, Technical Note D.2., World Bank (2003).
[24] Fitzhugh, T. and Richter, B. "Quenching Urban Thirst: Growing Cities and Their Impacts on Freshwater Ecosystems," **BioScience**, 54, 8 (August 2004). Some of the Millennium Ecosystem Assessment scenarios show increased levels of nitrogen pollution of water and coastal systems as a result of drivers including urbanization. Millennium Ecosystem Assessment, **Ecosystems and Human Well-being: Synthesis**, Island Press (2005).
[25] United Nations Population Division (UNPD), **World Urbanization Prospects**, UN Population Division (2002).
[26] http://www.wmo.ch/web/homs/documents/english/icwedece.html
[27] http://www.un.org/esa/sustdev/documents/agenda21/english/agenda21chapter18.htm
[28] http://www.un.org/esa/sustdev/documents/agenda21/english/agenda21chapter18.htm

account the functioning of aquatic ecosystems...."[29] However, the importance of maintaining adequate hydrologic flows are mentioned only once in all of Chapter 18.

Prior to and after the 1992 Earth Summit, international and regional efforts have aimed at advancing water management that protects ecosystems, including contracting party meetings for the Ramsar Convention on Wetlands and the Convention on Biological Diversity. The Convention on the Non-Navigational Uses of International Watercourses, which was adopted by the United Nations General Assembly in 1997 but has not yet entered into force, states that it is a duty for governments to conduct environmental impact assessments to evaluate whether and to what extent a particular project or activity will adversely impact the environment and to consider alternative projects and strategies for achieving the same objective. In 1993, the United Nations Economic Commission for Europe developed guidelines on the ecosystem approach in water management.[30] Documents like the World Water Action Report of 2002 and the Vision for Water and Nature[31] also emphasize the importance of conserving freshwater ecosystems.

The 13th Session of the Commission on Sustainable Development concluded with recognition of the role of ecosystems in water management;[32] however, throughout the negotiations there was bitter debate with a large segment of countries arguing for removal of any reference to ecosystems. In essence, what appears to be developing is a divide between those advocating for water supply and sanitation and those emphasizing an ecosystem-based approach to decision-making and management. A more socially-acceptable approach would be to seek ways to integrate human values associated with healthy ecosystems with other water resource values associated with agriculture, cities, and industries.[33] Even the United Nations Millennium

---

[29] http://www.un.org/esa/sustdev/documents/agenda21/english/agenda21chapter18.htm

[30] http://www.unece.org/env/documents/2004/wat/sem.4/mp.wat.sem.4.2004.4e.pdf

[31] IUCN, **Vision for Water and Nature: A World Strategy for Conservation and Sustainable Management of Water Resources in the 21st Century** (2000).

[32] "(v) Enhancing the sustainability of ecosystems that provide essential resources and services for human well being and economic activity in water-related decisionmaking;" "(ix) Protecting and rehabilitating catchment areas for regulating water flows and improving water quality, taking into account the critical role of ecosystems;" http://www.un.org/esa/sustdev/csd/csd13/csd13_decision_unedited.pdf

[33] Richter, BD, et. al, "Ecologically sustainable water management: managing river flows for ecological integrity," **Ecological Applications**, 13:206–224 (2003).

Declaration Goal #7 recognizes the need to integrate sustainable development principles into all policies and programmes to reduce the loss of environmental resources.[34] The challenge, however, is to put global policies into practice.[35]

5. Recognizing the Limits of Freshwater Ecosystems

To meet growing human needs for water, electricity, food, and flood control, humans have dammed and diverted waterways in the last century to the detriment of riverine ecosystems and related habitats (see Fig. 1). Over half of the 292 large river systems around the globe are highly or moderately fragmented by dam structures.[36] Dams impact on the hydrological cycle by changing the flow, temperature, and nutrient content of waterways (see Fig. 2). Dams of all sizes change the basic appearance of rivers and streams, while larger dams pose a barrier to species dispersal.[37] Not only do big dams impede river flow, greatly altering natural stream flow patterns and impoverishing habitats, many of them also reduce the amount of water that eventually reaches downstream areas, especially deltas and their associated wetlands.[38] Dams also reduce sediment deposition downstream; the less sediment-laden waters increase scour potential downstream as well as limit the amount of sediments (and nutrients) in deltas and marshes, for example, which are important to ecosystem regeneration and stability. Fish catches of native species often fall precipitously both upstream and downstream of large dams.[39] Dams can also affect groundwater aquifers with resultant impacts on downstream wetlands as a result of reduced flows.

---

[34] http://www.un.org/millenniumgoals/#; "To stop the unsustainable exploitation of water resources by developing water management strategies at the regional, national and local levels, which promote both equitable access and adequate supplies." http://www.un.org/millennium/declaration/ares552e.pdf

[35] As Malin Falkenmark points out, another challenge is to shift the focus away from the effects of water resources development on the environment but rather on assessing environmental impacts of water resources management strategies. Falkenmark, M., **Water Management and Ecosystems: Living with Change**, TEC Paper #9, GWP (2003).

[36] Nilsson, C., et al., "Fragmentation and Flow Regulation of the World's Large River Systems," **Science** 308: 405–408 (2005).

[37] World Commission on Dams, **Dams and Development: A New Framework for Decision-Making**, Earthscan Publications (2000).

[38] Hinrichsen, D., et. al., op. cit. 2003.

[39] World Commission on Dams, op. cit. 2000.

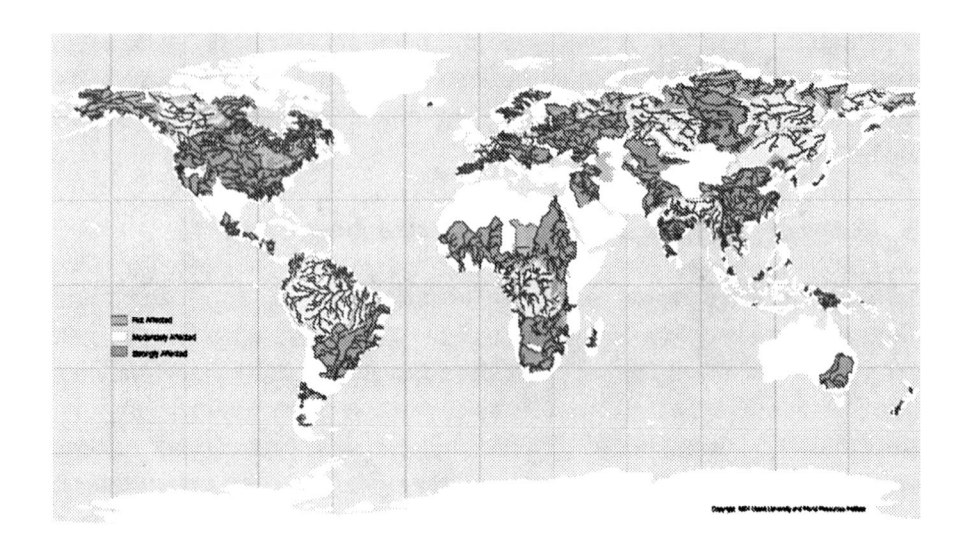

Figure 1. River Fragmentation and Flow Regulation

Source: Nilsson, C., C. A. Reidy, M. Dynesius, and C. Revenga. 2005. "Fragmentation and Flow Regulation of the World's Large River Systems" *Science*, 308, 5720, 405–408.

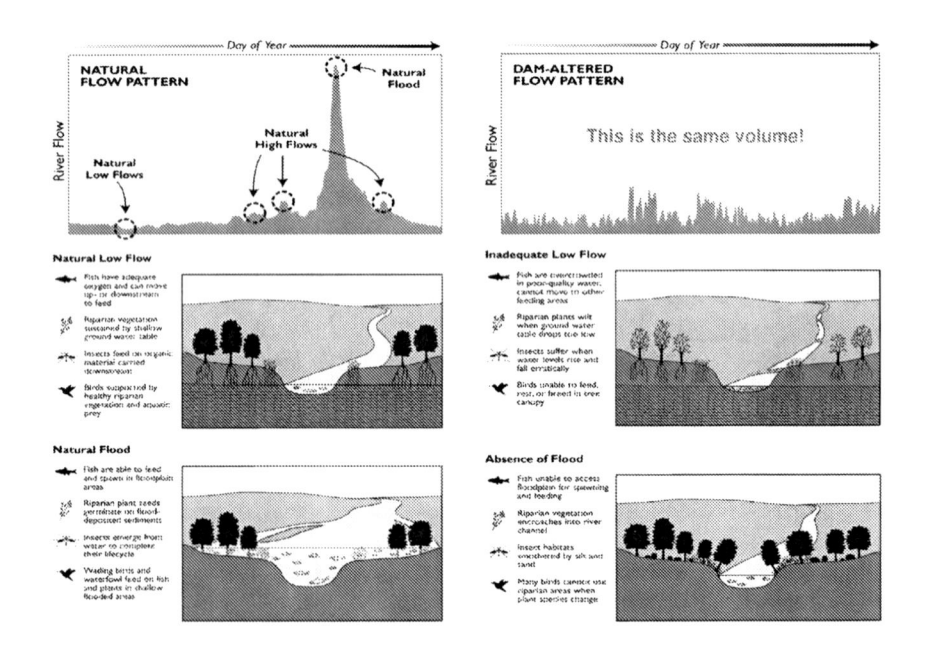

Figure 2. Impacts of Dams on River Flow

Source: Postel, S. and Richter, B., Rivers for Life: Managing Water for People and Nature, Island Press (2003).

Recognizing that human societies depend upon and receive valuable benefits from healthy ecosystems, Postel and Richter have suggested that the first priority in any freshwater allocation scheme should be to make an "ecosystem support allocation."[40] This allocation should be designed to ensure that ecosystems receive the quantity, quality, and timing of freshwater flows or inflows needed to safeguard the health and functioning of river systems and estuaries. This approach places a limit on the degree to which society can alter natural river flows or inflows to estuaries. Postel and Richter have called this limit the "sustainability boundary." Rather than freshwater and estuarine ecosystems getting whatever water happens to be left over after human demands are met, they receive what they need to remain healthy. As depicted in Fig. 3, modification of river flows for economic purposes expands over time, but only up to the sustainability boundary.

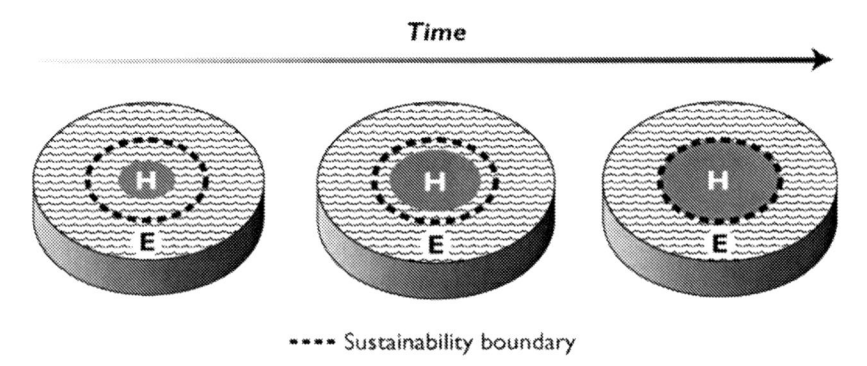

Figure 3. The Sustainability Boundary Concept

Source: Postel, S. and Richter, B., Rivers for Life: Managing Water for People and Nature, Island Press (2003).

In the diagram, human uses of water (H) can increase over time but only up to the sustainability boundary. At that point, new water demands must be met through conservation, improvements in water productivity, and reallocation of water among users. By limiting human impacts and allocating enough water for ecosystem support (E) society derives optimal benefits from healthy catchment and estuarine systems in a sustainable manner.

---

[40] Postel, S. and Richter, B., **Rivers for Life: Managing Water for People and Nature**, Island Press (2003).

Over the past few decades, dam designs and turbine technology have advanced in allowing for fish passage (including prevention from impingement and turbine entrainment) and maintenance of fish habitat. More is known on reservoir fluctuations and the effects of fluctuating flows on aquatic species and wetlands. Mitigation measures exist to cope with such issues as dissolved oxygen. Moreover, physical and numerical modeling techniques and laboratory and field evaluations allow for improved understanding to contribute to management decisions. However, much more needs to be done, particularly in integrating environmental flow requirements into dam design and operations.

As noted above, environmental flows are the amounts of freshwater flowing in rivers and streams and into bays and estuaries that are needed to support healthy, diverse populations of fish and other aquatic life. The definition of environmental flows may capture practices described by other terminology common to particular countries or in specific geographies such as ecological flows (China), ecological reserve (some parts of Africa), and instream flow protection (USA). At the core of the concept is the understanding that variable flow conditions are required to enable aquatic species to complete their life cycles.

6. Tools for Addressing Environmental Flows

As described above, the health of freshwater ecosystems depends on some amount of water remaining in the system to support the needs of both humans and biodiversity.[41] Ecosystems vary, but aquatic species depend on an appropriate sequence and timing of minimum flows, normal flows, and peak flows for their survival.[42] The productivity levels of deltas and estuaries, which arguably rival that of wetlands, also depend on the timing and volume of freshwater inflows to support critical habitats such as mangroves as well as fishery stocks.[43]

In the last few decades, environmental flow science has progressed considerably. However, guidance on the linkages between hydrology

---

[41] Dyson, M. et. al. (Ed.), **Flow: The Essentials of Environmental Flows**, IUCN (2003).
[42] Water quality is also central to the viability of ecosystems. However, this paper focuses on water quantity issues for ecosystem health.
[43] Postel, S. and Richter, B., op. cit. 2003.

and ecology still remain limited.[44] River ecosystem restoration projects across the globe seek return to some degree of the "natural" state of rivers.[45] Over 850 river flow restoration projects are being implemented in more than 50 countries.[46]

Figure 4 is a conceptual illustration of the objective of a process to develop environmental flow recommendations – determining the desired flow of water in a natural river or lake that sustains healthy ecosystems and the goods and services that humans derive from them. The light gray area represents the natural flows for a particular river. The dark gray represents the part of the flow regime that must be protected to conserve ecosystem health.

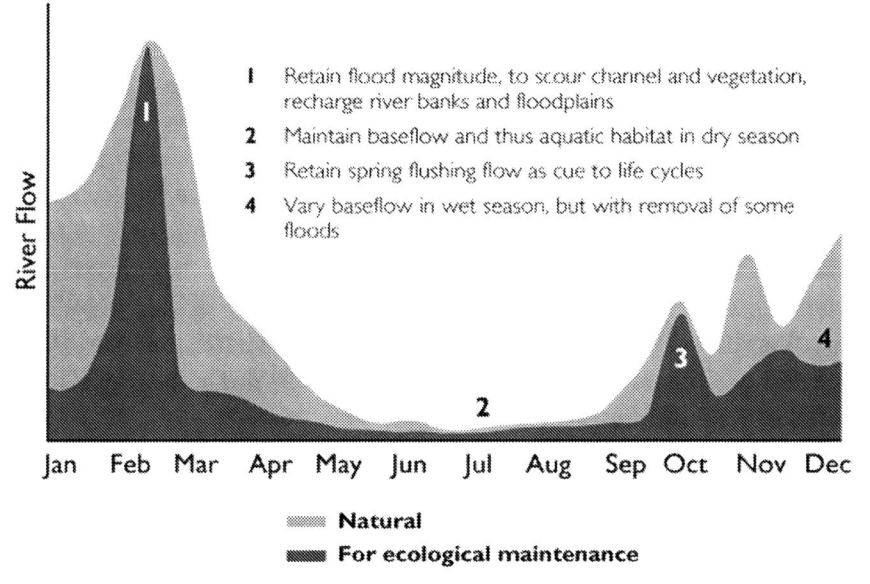

Figure 4. Conceptual Illustration of River Flow Development

Source: Postel, S. and Richter, B., **Rivers for Life: Managing Water for People and Nature**, Island Press (2003).

Despite improved methods for scientifically defining a river ecosystem's flow needs, water management processes continue to fail to account for scientific understanding of freshwater ecosystems.[47]

---

[44] Falkenmark, M., op. cit. 2003.
[45] Postel, S. and Richter, B., op. cit. 2003.
[46] The Nature Conservancy, **Flow Restoration Database**, www.freshwaters.org/tools.
[47] Richter, B. et. al. **A Collaborative and Adaptive Process for Developing Environmental Flow Recommendations**, River Research and Applications 22: 297–318 (2006).

Furthermore, in many countries, the existing hydro-meteorological networks are poorly maintained and have significant data gaps.[48] Environmental flow policy and science is most advanced in South Africa, the United States, the European Union, and Australia.[49] However, countries across Latin America are working to advance the study of hydrology and integrate it into water management processes, through the International Hydrological Programme (IHP) of UNESCO for Latin America and the Caribbean and its related programs such as the Programme on Hydrology for the Environment, Life and Policy.[50] One challenge for environmental flow science is the lack of comprehensive information on habitat and biota in the river systems.

Another challenge, one that applies to both developed and developing countries, is the complexity of developing environmental flow recommendations that are aligned with social goals, particularly in ways that involve all stakeholders in deciding upon the health of a country's rivers. New processes are being offered by scientists and water managers to address the need for comprehensive scientific assessments of ecosystem needs, while providing flexibility to address social needs and desires along the development spectrum. For instance, the DRIFT method developed in South Africa explicitly identifies the different degrees of ecological health that would be expected as existing flow conditions in a river are either increasing altered or restored.[51]

Environmentally sustainable water management is one approach being applied in the reoperation of non-FERC (Federal Energy Regulatory Commission) dams in several sites in the United States that offers the potential to support dam decision-making processes in other countries as well. ESWM takes into account not simply the needs of an individual fish species but also the flow requirements for riparian vegetation, floodplain wetlands, downstream estuaries, wildlife species, and other associated ecosystem services and products.[52] This framework – known as ESWM – meets human water

---

[48] Millennium Ecosystem Assessment, **Conditions & Trends Assessment: Freshwater**, op. cit. 2005.
[49] Postel, S. and Richter, B., op. cit. 2003.
[50] http://www.unesco.org.uy/phi/
[51] Postel, S. and Richter, B., op. cit. 2003.
[52] Richter, B., et. al. "A Framework for Ecologically Sustainable Water Management," **Hydro Review** (July 2005).

needs by storing, diverting and releasing water in a manner that either sustains or restores a river's ecological integrity. ESWM is a six-step framework that involves: (1) Developing estimates of ecosystem flow requirements; (2) Accounting for current and future human uses of water; (3) Identifying incompatibilities between human and ecosystem needs; (4) Using a collaborative approach to find solutions to resolve the incompatibilities; (5) Conducting water management experiments to reduce uncertainties that frustrate efforts to integrate human and ecosystem needs; and (6) Designing and implementing an adaptive management plan (see Fig. 5).[53] The ESWM framework understands the flow necessary to sustain or restore the integrity of a river, assesses human influence on water flow, and identifies areas of incompatibility between people and nature to provide a foundation for next steps.

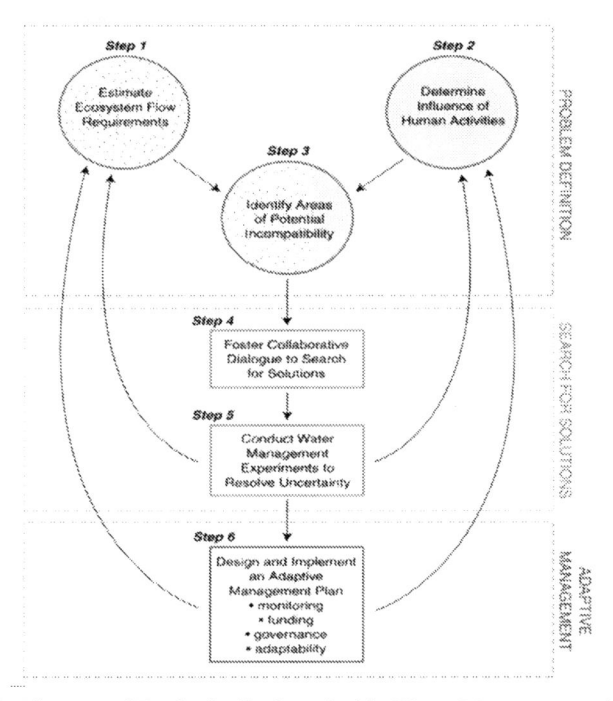

Figure 5. Diagram of Ecologically Sustainable Water Management Approach

Source: Richter, BD, et. al, "Ecologically sustainable water management: managing river flows for ecological integrity," Ecological Applications, 13:206–224 (2003).

---

[53] Richter, BD, et. al, "Ecologically sustainable water management: managing river flows for ecological integrity," **Ecological Applications,** 13:206–224 (2003).

Over the past 10 years, the Nature Conservancy has been working on developing the ESWM approach and applying it to dams in the United States and more recently to dams outside of the USA Since 2002, the Conservancy has been partnering with the US Army Corps of Engineers to apply ESWM in the Savannah River, among other demonstration sites. Dividing the States of Georgia and South Carolina, the Savannah River is impacted by three multipurpose dams authorized for hydropower generation – the Hartwell, Russell and Thurmond dams. In the Savannah process, the Nature Conservancy worked with the US Army Corps of Engineers and a multi-disciplinary team of scientists, through a five-step process, to develop environmental flow recommendations for the river. This included: (1) an orientation meeting (2) a literature review and the development of a summary of existing knowledge about flow-dependent biota and ecological processes of concern; (3) a workshop which involved development of ecological objectives and initial flow recommendations, and identified key information gaps; (4) implementation of the flow recommendations on a trial basis to test hypotheses and reduce uncertainties; and (5) monitoring system response and carrying out additional research. A range of recommended flows were developed for the low flows in each month, high flow pulses throughout the year, and floods with targeted inter-annual frequencies.[54]

In the Savannah River, the ESWM process ultimately resulted in the flow recommendations being incorporated into a comprehensive river basin planning process conducted by the US Army Corps of Engineers and used to initiate the adaptive management of Thurmond Dam. Figure 6 depicts one realization of the environmental flow prescription process in the Savannah River where each element was tied to an ecological hypothesis to help managers determine how to proceed in ensuring the variability needed to sustain a healthy ecosystem.

In general, outcomes from application of an ESWM approach may include developing collaborative solutions to resolve incompatibilities, conducting experiments to test new approaches and designing and

---

[54] Richter, B. et. al., op. cit. 2006.

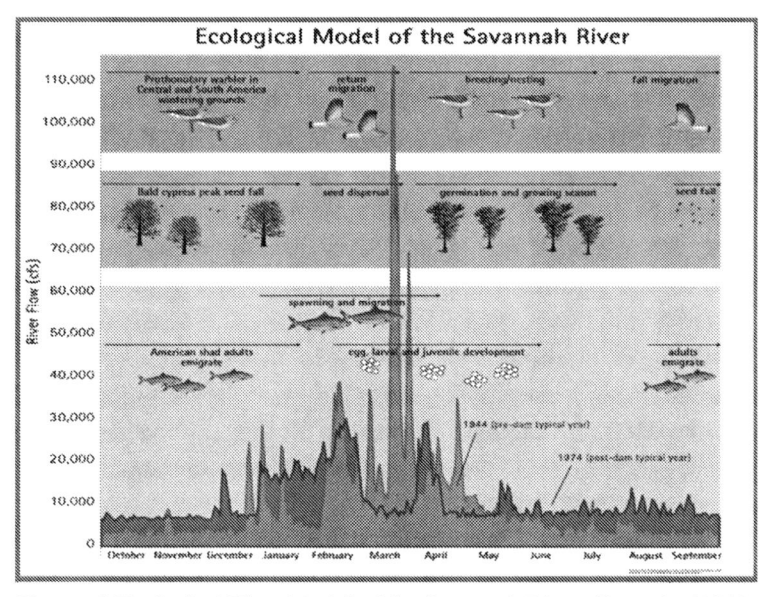

Figure 6. Ecological Flow Model of the Savannah River (Georgia, USA)
Source: The Nature Conservancy.

implementing an adaptive management plan to improve water management for the long term. The recent Millennium Ecosystem Assessment clearly points out the advantages of adaptive management.[55] Based on current knowledge, some actions may be taken to restore environmental flows. In other cases, additional analysis may be needed, over periods of months and years. Re-allocation of water uses (for example among agricultural, industrial, municipal, and environmental uses) may be required and this can take substantial time due to changes needed in laws, incentive structures, and behaviors. Throughout this period, adaptive management is needed as improved monitoring and data collection systems are put into place, more information is obtained, and scientists, water managers, and stakeholders work toward consensus through facilitated dialogues and learning network approaches. In the ESWM approach, Steps 3–5 are repeated indefinitely to enable iterative refinement of environmental flow recommendations.[56]

---

[55] Millennium Ecosystem Assessment, op. cit. 2005.
[56] Richter, B. et. al., op. cit. 2006.

7. Future Methods for Improving Water Management

River basins cross political boundaries (whether national or sub-national) and cut across many users with competing demands for the precious resource of water. River Basin Management is a key challenge and opportunity for sustainable water resource management. Countries are establishing river basin organizations as a tool for improved water management, moving toward Integrated River Basin Management and Integrated Water Resource Management. While some river basin organizations have been functioning for some time, many are just being established or function only on paper. Plans and institutions will only succeed, however, if the policies and programs exist to bring about real changes in ensuring water and development needs are met at local levels.

The Mekong River Commission's Environment Program includes an Environmental Flow Component with an environmental flow assessment underway. But much more work needs to be done, particularly on sharing lessons learned from environmental flow assessments carried out in a transboundary basin.

As noted above, despite vast improvements in information, insufficient data exist on water quantity and quality requirements of freshwater ecosystems in many river basins. Thus, many rivers never have environmental flow requirements developed because of lack of money or time to collect and analyze the data. This does not mean that environmental flows should not be considered in river basin management. New approaches are being developed that can help guide river basin organizations, governments, industry, water managers, and communities to determine how to ensure healthy freshwater ecosystems and sustainable livelihoods for those dependent on the river systems. A new approach referred to as the Limits of Hydrologic Alteration (LOHA) is a way to promote water management strategies that are tailored to both ecological considerations and water quantity limitations.[57] Using information about desired river health goals and existing hydrologic alteration, rivers are assigned specific hydrologic alteration limits. This approach helps direct limited resources to areas of greatest ecological health or restoration potential.

---

[57] Richter BD, Apse CD, and Warner AT. "Beyond Tennant: A Call for a New Approach in Environmental Flow Science." In Review, River Research and Applications.

River basin organizations offer a vehicle through which a range of partners can work together reconcile the short-term emphasis on development with the long-term view of creating compatible biodiversity conservation and sustainable human use within a river basin and different river cultures.

8. Conclusion

Water issues are often discussed as finding a balance between human and nature. This unfortunately continues the misconception that maintaining water in the environment is somehow not to the benefit of humans. Much work remains to be done to find new and creative ways to slow and reverse the rapid loss of ecosystems, particularly in the next half century when an additional three billion people will need to share in the use of our limited natural resources. Political decisions often are based on shortsighted goals of reelection, career advancement, and party positioning. That said, the Millennium Ecosystem Assessment raises the alarm that freshwater ecosystems are being degraded at a rapid rate and that this may undermine future well-being. The challenge will be to link the passion for ecosystem conservation to compassion for human well-being. This will require using what we already know about policy responses, testing new and holistic approaches, strengthening governance systems, letting go of the tendency for "territoriality," and finding ways to bring about multiple benefits with policy actions – ones that recognize the vital links among growing populations, escalating resource demands, shrinking supplies and basic human goals of health and economic security.

THE AUTHOR

Karin M. Krchnak graduated in Political Science from Duke University and received her Juris Doctor from the University of Maryland, School of Law. At TNC, she is part of a core team working to reduce the environmental footprint of hydropower projects. For over 15 years, she has worked for consulting firms and nonprofits to improve policies and procedures related to environmental management and resource conservation worldwide. She serves on the Steering Committee of the Global Water Partnership.

TRANSBOUNDARY AQUIFERS AS KEY COMPONENT OF INTEGRATED WATER RESOURCE MANAGEMENT IN CENTRAL ASIA

KEN HOWARD AND ANNE GRIFFITH
Groundwater Research Group, University of Toronto at Scarborough,
1265 Military Trail, Toronto, Ontario M1C 1A4, Canada

Abstract: Since the early 1960s, mismanagement of water resources has plagued the Aral Sea Basin. The problem became more complex in 1991 when the Soviet Union collapsed and the Aral Sea Basin became a transboundary water resource. Overnight, the development of sustainable and equitable water management practices became the shared responsibility of five sovereign nations each with conflicting needs, goals and priorities. Solutions presented thus far by scientists, international institutions, and governments to ameliorate the problems in the Aral Sea Basin have failed. To a large extent, this can be explained by political tensions in the region and the difficulties of reaching consensus on appropriate and effective management strategies. However, serious questions must also be raised whether the solutions proposed – solutions that focus almost exclusively on the equitable allocation of surface water flows – are destined to fail unless the entire resource including groundwater is considered. It is argued that any strategy for the management of water resources in the Aral Sea Basin must have appropriate regard for the vital and very significant role groundwater plays in the overall water budget. Globally, over 98% of all fresh, accessible water is found in aquifers, but groundwater is frequently neglected in resource planning because of an "out-of-sight, out-of-mind" mentality. Data suggest that a number of important transboundary aquifers occur in the Central Asia region, and that there is a need, not only to manage these internationally-shared resources in close collaboration with neighboring countries, but to manage these aquifers conjunctively with surface water as part of a fully integrated water resource management strategy. Although the transboundary nature of the region's water resources makes equitable and sustainable management a formidable challenge, the task is surmountable with robust scientific data and if proper monitoring techniques, institutional arrangements, research collaboration, and programs that fully acknowledge the vital role of regional transboundary aquifers are implemented.

Keywords: groundwater, aquifers, Integrated Water Management (IWM), equitable and sustainable management practices

1. Background – Water and Environmental Issues in Central Asia

Serious mismanagement of water resources is classically exemplified by the fate of the Aral Sea in Central Asia. An ambitious attempt to use surface water to spur economic growth has led to severe environmental degradation. The Aral Sea (Fig. 1) was once the world's fourth largest inland sea, and supported a fishing industry that employed 60,000 people. Today, the Aral Sea has been reduced to one tenth of its original size and the fishing industry has been decimated. The settlement of Muynak, a coastal fishing village is now 100km from the nearest open water and the fertile lands that once surrounded the sea have become severely degraded by chemicals blowing off the dried sea bed. Halophytes have emerged as the dominant form of vegetation (Saiko and Zonn, 2000).The Aral Sea region has become an inhospitable desert wilderness with one of the highest infant mortality rates in the world.

The problems began in the 1960's during the Soviet Union era when two of the basin's most important rivers, the Amu Darya and the Syr Darya, were harnessed in the mountains to generate hydroelectric energy, and diverted in their lower reaches to irrigate vast expanses of semiarid plains for the production of cotton. River flows to the Aral Sea were reduced substantially and over a period of 30 years, the level of the sea declined by 17 m.

Figure 1. The Aral Sea Basin of Central Asia

The massive network of irrigation canals were plagued by inefficiencies with the majority of farm land receiving a supply of water that was simply inadequate for maximising crop productivity (McCray, 1998). It has been estimated that less than half the irrigated water ever reached the crops. Large-scale irrigation has compounded the serious problems facing the region by causing a rise in groundwater tables and a loss of soil fertility due to salinisation, and promoting downstream dispersion of agricultural chemicals. With up to 90% of people in parts of the Central Asia region dependent upon agriculture for employment (UNDP et al. 2005) the situation is dire with potentially devastating socioeconomic consequences. The region is already plagued with falling cotton production in an economic downturn that, thus far, shows no signs of abating. While solutions to the region's social, economic and environmental crisis are clearly complex and inevitably long-term, none can be successful unless they address the root cause of the problem i.e., water resource management practices that failed to respect the principles of equity and sustainnability.

The situation assumed a greater level of complexity in 1991 when the Soviet Union collapsed and Kazakhstan, Kyrgyzstan, Uzbekistan, Tajikistan, and Turkmenistan (Fig.1) inherited an ailing and unsustainable system of water utilization in desperate need of repair. Overnight, a collective scheme of resource utilization, orchestrated and policed by a single top-down system of governance (Djalalov, 2003; McKinney 2004) became the responsibility of five sovereign nations. The Aral Sea Basin had become a shared "transboundary" water resource and the development of sustainable and equitable water management practices became a shared responsibility.

Inevitably, problems arose due to the conflicting needs, goals and priorities of the custodial nations. These divergent interests escalated the abuse and mismanagement of the available resources and have led to serious mistrust and tension in the region. An annual cycle of disputes has developed between the economically weaker headwater nations, Kyrgyzstan and Tajikistan whose interests lie mainly in farming and power generation, and the heavy downstream consumers Turkmenistan, Kazakhstan, and Uzbekistan. Although collaborating governance bodies have been established, the unsustainable use of water has continued unabated – as have the negative consequences.

Solutions presented thus far by scientists, international institutions, and governments to ameliorate the problems in the Aral Sea Basin have repeatedly failed. Some suggest that the basin is non-restorable (McKinney, 2004). To a large extent repeated failures can be explained by political tensions in the region and the difficulties of reaching agreement on appropriate management strategies. However, serious questions must also be raised whether the solutions proposed – solutions that focus almost exclusively on the equitable allocation of surface water flows – are doomed to failure unless the entire resource, including groundwater, is considered.

In this paper, we argue that any strategy for the management of water resources in the Aral Sea Basin must have due regard for the vital and very significant role groundwater plays in the overall water budget. Over 98% of all potable fresh water is to be found in aquifers, but groundwater is frequently neglected because of an "out-of-sight, out-of-mind" mentality. Data suggest that a number of important transboundary aquifers can be found in the Central Asia region and that not only is there a need to manage these internationally-shared resources in cooperation with neighboring countries but to manage these aquifers conjunctively with surface water as part of a fully integrated water resource management strategy. Although the transboundary nature of the Aral Sea Basin's water resources system makes equitable and sustainable management a formidable task, the challenge is surmountable if proper monitoring techniques, institutional arrangements, research collaboration, and programs that fully include groundwater and regional aquifers are implemented.

2. Global Recognition for Transboundary Aquifers

Although the hydrogeological community has been aware of the management complexities of transboundary aquifers since the early 1970s, very little was done to promote awareness of the issue in the international arena. The major focus was, and largely remains, on transboundary surface waters through international agreements that set quotas for water withdrawals and establish protocols for monitoring contamination. Worldwide, nearly 150 countries have part of their territory in a transboundary river basin and over 250 major rivers cross international boundaries, many several times. Until recently, there were over 400 treaties around the world dealing with surface water, and only 62 of those made any mention of aquifers. Of these 62 treaties, eight dealt with groundwater quantity and six addressed groundwater quality.

Global recognition for the function and significance of transboundary aquifers is largely attributable to the pioneering work of Dr. Shammy Puri, the founding Chair of the International Association of Hydrogeologists (IAH) Commission on Transboundary Aquifers. The Commission formed in 1997 and subsequently sought cooperation from UNESCO, FAO and the UNECE in the development of the ISARM (Internationally Shared Aquifer Resources Management) Program (Puri, 2001; Puri and Aureli, 2005). ISARM was later incorporated within the IHP–VI Program, and currently benefits from the support of not only the IAH National Committees but also the National Hydrological Committees of all the member states of the IHP (149 countries). It is estimated that internationally shared aquifers supply 40% of the world's population with water.

Potential issues related to transboundary groundwater flow are illustrated in Fig. 2. It is not unusual for an aquifer in one country to have received much of its recharge (and any entrained contaminants) from a neighboring country and, in some cases, transboundary flow of groundwater can take place in opposite directions at different depths. Eckstein and Eckstein (2005) postulate six basic conceptual models of shared aquifer systems which they propose can provide help in assessing the applicability and scientific reliability of existing and proposed rules governing transboundary groundwater resources. These models are shown in Fig. 3.

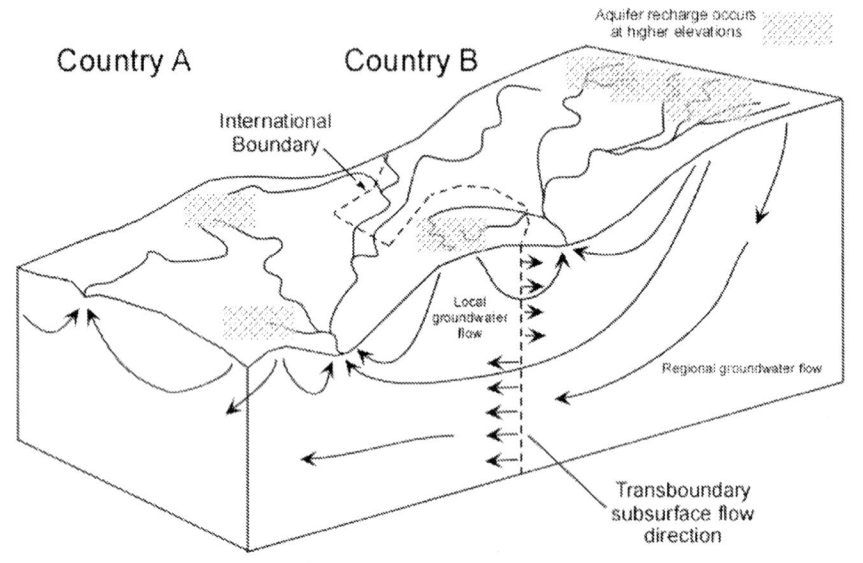

Figure 2. Potential issues related to groundwater flow in transboundary aquifers (modified after Puri (2001) and Howard (2004)). A transboundary aquifer is an aquifer that underlies, or whose water flows beneath, two or more political jurisdictions and has utility in each jurisdiction (e.g., provides baseflow to streams, supports groundwater dependent ecosystems or is pumped for supply).

Despite growing awareness for the importance of internationally-shared aquifers, the challenge of developing practical and enforceable legal agreements (or "treaties") between neighboring states has proved extremely difficult. Transboundary rivers can be managed by a relatively simple agreement that allocates water to partner states as a percentage of total flow, something that is easily monitored for compliance. However, this type of approach is clearly inappropriate for transboundary aquifers which are three-dimensional, relatively high storage systems with low, sometimes negligible replenishment rates and a very slow response to external stresses such as pollution and climate change. Shared aquifer systems pose special problems as resource management decisions made in one jurisdiction can have serious long- term impacts on the value of the resource in the other. A major challenge is to ensure internationally-shared aquifers are managed cooperatively, with competing demands for domestic supply, sanitation, agriculture, industry, mining and water-dependent ecosystems fully addressed and mutually agreed. A complementary but equally difficult challenge will be the development of appropriate enforcement and compliance indicators that will ensure all parties observe the terms of any resource management agreement.

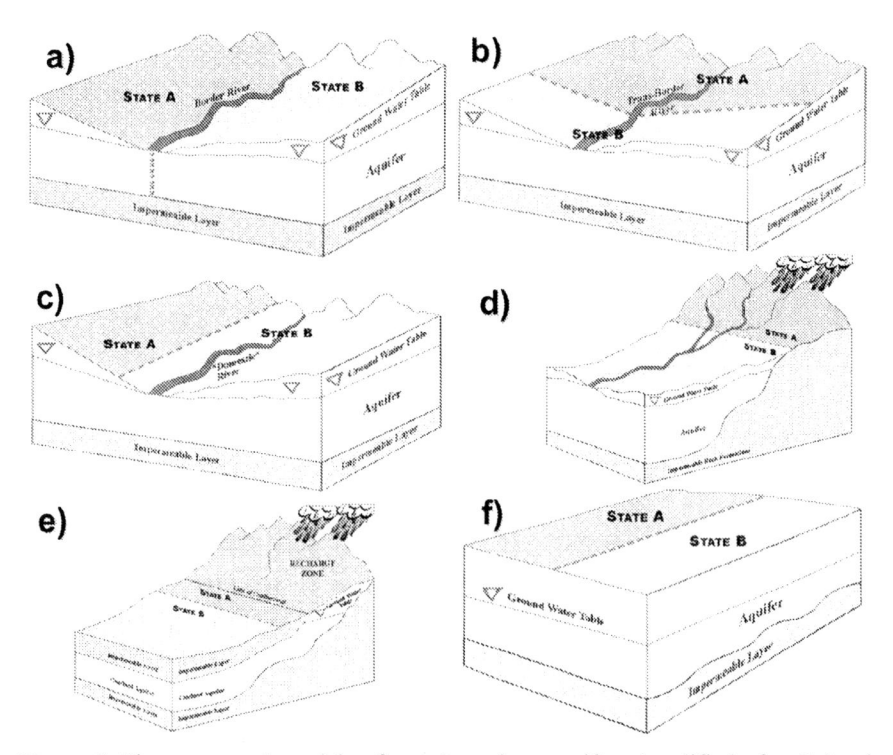

Figure 3. Six conceptual models of transboundary aquifers (modified after Eckstein and Eckstein (2005)). (a) Unconfined aquifer, hydraulically linked to a river which flows along the border; (b) Unconfined transboundary aquifer, hydraulically linked to a transboundary river; (c) Unconfined aquifer, hydraulically linked to a river which is completely contained within one territory; (d) Unconfined aquifer that occurs completely within one territory that is hydraulically linked to a river which is transboundary; (e) Confined transboundary aquifer that occurs within a single territory that is unconnected to a river; (f) Unconfined transboundary aquifer that is unrelated to any surface water feature and may even be disconnected from the hydrologic system (i.e., effectively zero recharge).

Most of the policy and law makers who are responsible for designing programs or developing treaties that consider transboundary aquifers utilize concepts based on an understanding of surface water flow and fail to appreciate the special nature of subsurface water and its very slow response to changes in management strategy. The UN's International Law Commission recently produced treaty drafts outlining methods of governance for transboundary aquifers, and providing rules for utilization, monitoring and activities that could affect neighboring states (Yamada, 2005). However, a number of issues were left deliberately vague, requiring, for example, that aquifer states simply "agree on harmonized standards and methodology for monitoring" and avoiding the fact that monitoring programs, as they currently exist may well prove ineffective for their intended purpose of ensuring equitable and sustainable use of the resource.

3. Transboundary Aquifers in Central Asia

One of the most important developments in recent times has been "WHYMAP" (The Worldwide Hydrogeological Mapping and Assessment Programme), an initiative of the German Federal Institute for Geosciences and Natural Resources (BGR) supported by a consortium that includes UNESCO and IAH among others. At the World Water Forum in March, 2006, WHYMAP released "Groundwater Resources of the World – Transboundary Aquifer Systems 1:50,000,000" with the aim of further promoting awareness for internationally shared aquifers. Part of the World Transboundary Aquifer Systems Map (modified after BGR (2006)) is shown in Fig. 4. Important transboundary aquifers are shown to occur throughout Central Asia although an exhaustive study of these systems has yet to be completed.

The WHYMAP Central Asia region is reproduced in greater detail in Fig. 5. It shows major groundwater basins associated with both the Amu Darya and the Syr Darya rivers and these aquifer systems are inevitably connected to the river systems. Significantly, the aquifer associated with the Amu Darya River also extends beneath the Aral Sea. In effect, the aquifers and rivers of Central Asia form a single water resource for the region and urgently merit "conjunctive" management as an integrated system.

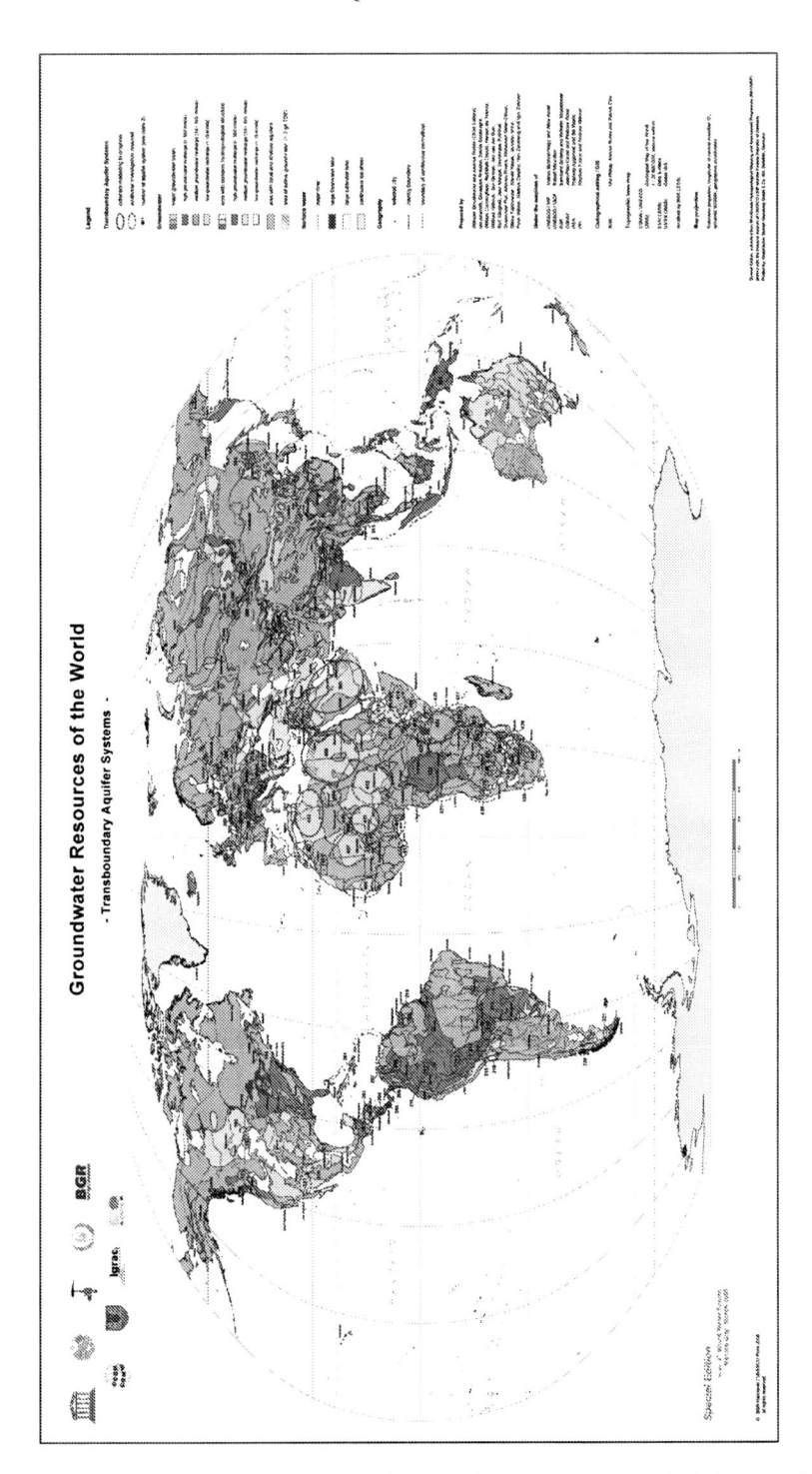

Figure 4. The World Transboundary Aquifer Systems Map modified after BGR (2006).

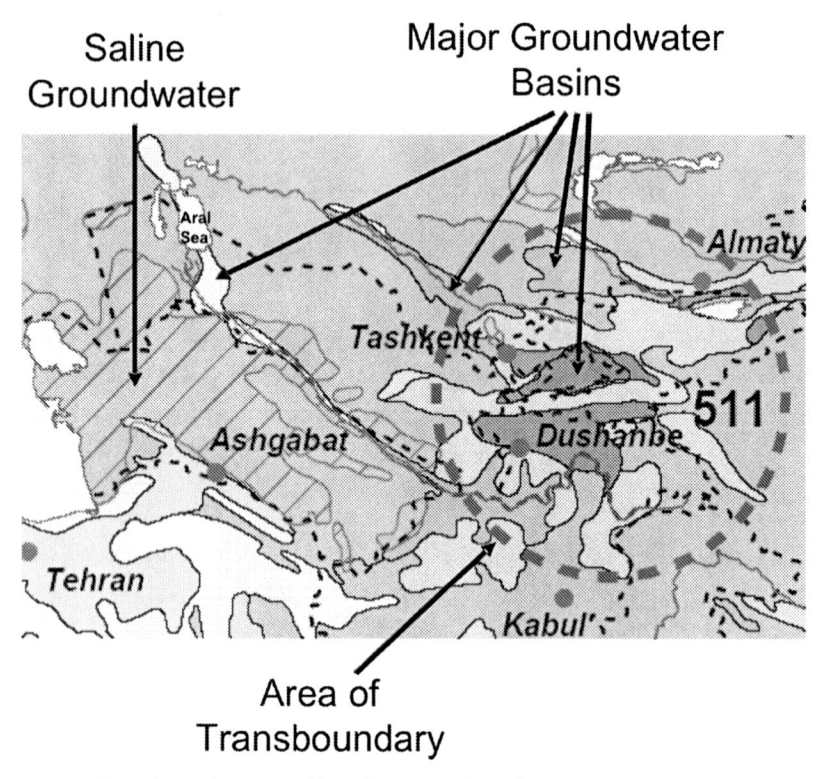

Figure 5. Transboundary Aquifers in Central Asia reproduced from the World [Transboundary Aquifer Systems Map (BGR 2006)]

4. Status of Transboundary Aquifer Management in Central Asia

Much of Central Asia comprises the Turanian Plain which includes the Kyzyl–Kum and Kara–Kum deserts (Bjorklund, 1998). The Tien Shan and Pamir mountains are to the southeast and are the main water source for both rivers and aquifers. Between the mountains and valleys are foothills and alluvial valleys where the main aquifers are found (McCray, 1998). The main source of recharge in this semiarid environment is glacial meltwater from the mountains. There is a negative correlation between the volume of water originating within a country and its available arable land. For example, Uzbekistan has the largest percentage of arable land whereas over 90% of its water originates from the mountains of Tajikistan, Afghanistan, and Turkmenistan. A generalized southeast to northwest cross section is shown in Fig. 6. Recharge occurs primarily in the elevated southeast where precipitation considerably exceeds evapotranspiration. Regional groundwater flow is generally northwest in the direction of the Aral Sea.

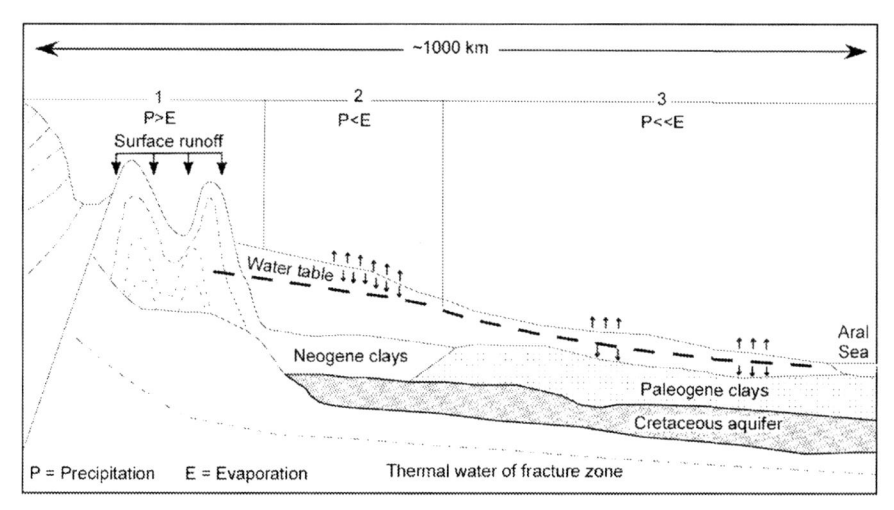

Figure 6. A generalized southeast to northwest cross section across Central Asia (modified after McCray, 1998)

Due to the economic importance of the oil and gas industry, most geological studies have focused on petroleum geology and there have been very few in depth scientific studies published in English on the region's hydrogeological setting. This has significantly hindered any advancement in the development of regional aquifer resource management plans. A few important studies have been completed in areas of agricultural, economic or political importance. For example:

- Morris et al. (2002 and 2005) studied the impacts of urbanization on the groundwaters of Bishkek (Fig. 7), the capital of Kyrgyzstan (population 900,000), which is entirely dependent on groundwater for domestic and industrial water supply. These studies demonstrate the massive reserves of groundwater stored in the mountain valleys where fluvioglacial/alluvial sediments form highly permeable aquifers (transmissivity up to 6000 m^2/day) hundreds of meters thick.

- Similarly, Pozdniakov and Shestakov (1998) studied an alluvial valley of the Kafernigan River, a tributary of the Amu Darya, featuring a 200–300m thick sequence of Quaternary sediments comprising interbedded lenses of sand, silt, clay, gravel, and loam. Finer sediments were found in the centre of the valley and graded to gravel at the edges. The transmissivity of the aquifer was estimated to be 2000m^2/day.

- Grebenyukov (2001) investigated the Ili River Basin in Kazakhstan and demonstrated a strong correlation between river flow and groundwater fluctuations in the unconfined aquifer. The aquifer comprised uniformly fine alluvial and proluvial Quaternary and Pliocene sands approximately 200–220m thick with a hydraulic conductivity of 6.1m/day.

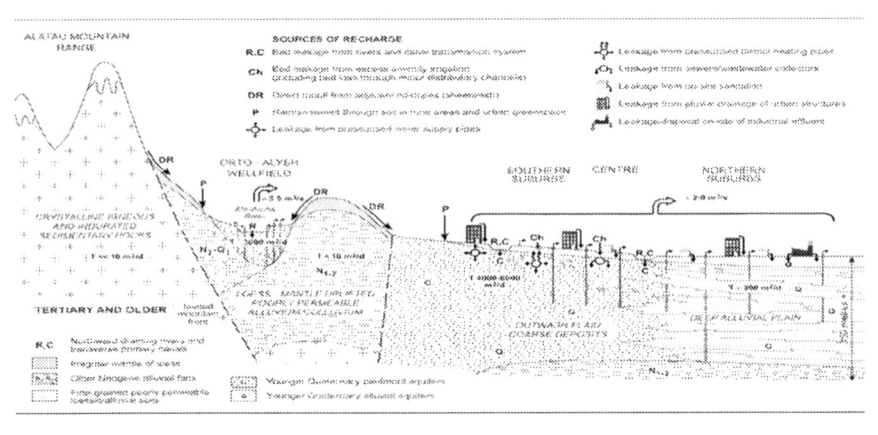

Figure 7. Cross-section of the aquifer underlying Bishkek, Kyrgyzstan (from Morris et al. 2002 and 2005)

Projects on the Amu Darya Basin and the Aral Sea are abundant (e.g., INTAS, World Bank, UNEP, NATO) but projects involving groundwater (e.g., defining aquifers, monitoring potentiometric levels and levels of contamination) are rare, piecemeal and do not capture the regional situation. For example:

- The USGS (Ulmishek, 2004) documented the presence of the Shatlyk aquifer, a thick water bearing sandstone of Lower Cretaceous age with lateral continuity over a vast area (shown on Fig. 6). In many areas, this aquifer is better known for its gas reserves than for its supply of good quality water. However, it will likely produce potable supplies of water in the foothills, and downgradient where the water is more saline, the water could be used to re-establish vegetation in and around the Aral Sea.

- Veselov et al. (2004) identified four aquifer complexes along the eastern edge of the Aral Sea in Kazakhstan:
  - Paleogene–Neogene–Quaternary sediments (3–5m thick);
  - Upper Turonian–Cenomanian sediments 5–15m thick) (Late Cretaceous);
  - Upper Albian–Cenomanian sediments (Middle Cretaceous);
  - Lower to Middle Jurassic sediments.
  - Benduhn and Renard (2004) documented three sources of groundwater close to the Aral Sea and suggested that flow to the Aral Sea was strongly influenced by the level of the sea and the pathways of groundwater flow. Gascoin and Renard (2005) calculated that the groundwater contribution to the Aral Sea to be 4 km^3/yr and without this flow component, the Aral Sea would be disappearing at a faster rate than it is currently. According to Jarsjo and Destouni (2004) groundwater represented 12% of the inflow into the Aral Sea in 1960 while today it is close to 100%.

Despite these examples, there is, overall, a noticeable paucity of information on the hydrogeology at the regional scale. The Central Asia Regional Water and Environmental Information System (CAREWEIS) estimates there are 339 aquifers in the region of which 30% are transboundary in nature, while the Asian Development Bank (2003) concludes that shared basins number 300. In terms of quantitative data, most relate to usable reserves and ignore the fact that in developing effective resource management strategies the temporal and spatial distribution of groundwater production are more important than the total aquifer yield.

The information that exists on the volume of groundwater available in Central Asia varies between sources but is generally consistent. Data from the Interstate Committee for Water Coordination (ICWC) and CAREWEIS are shown on Table 1 and do take into account water quality which can limit its use in some areas. Although the quantities shown are somewhat lower that the annual river flows of the Syr Darya (37.58 km^3) and Amu Darya (77.27 km^3) they ignore the huge volumes, likely orders of magnitude higher, that remain in aquifer storage and can be readily utilized as part of a resource management strategy.

TABLE 1. Regional estimates of groundwater availability (Interstate Committee for Water Coordination (ICWC) 2002 and Central Asia Regional Water and Environmental Information System (CAREWEIS) (data from current website).

Country	Estimated regional reserves km^3/year		Reserves approved for use km^3/year		Actual withdrawal 1999 km^3/year	
	ICWC	CAREWEIS	ICWC	CAREWEIS	ICWC	CAREWEIS
Kazakhstan	1.846	1.846	1.27	1.224	0.293	0.42
Kyrgyzstan	1.595	0.862	0.623	0.67	0.244	0.407
Tajikistan	18.7	6.650	6.02	2.200	2.294	0.990
Turkmenistan	3.36	3.360	1.22	1.220	0.457	0.457
Uzbekistan	18.455	18.455	7.796	7.796	7.749	7.749
Total Aral Sea Basin	43.486	31.173	16.938	13.110	11.037	10.023

Although the majority of groundwater extracted is used for irrigation (Table 2) very little groundwater is used for irrigation purposes compared to surface water. According to the UN Food and Agricultural Organization (FAO) groundwater accounts for only 5.6% of all the irrigated land in Central Asia. In fact, the most important use of groundwater at present is its contribution to the Aral Sea.

TABLE 2. Uses of groundwater in the economies of Central Asia (data from CAREWEIS website).

Country	Actual groundwater withdrawals km^3/year	By sector – km^3/yr					
		Drinking water supply	Industry	Irrigation	Vertical drainage	Pumping tests	Other
Kazakhstan	0.42	0.2	0.081	0	0	0	0.012
Kyrgyzstan	0.407	0.043	0.056	0.145	0	0	0
Tajikistan	0.990	0.485	0.2	0.428	0.018	0	0.06
Turkmenistan	0.457	0.21	0.036	0.15	0.06	0.001	0.112
Uzbekistan	7.749	3.369	0.715	2.156	1.349	0.12	0.04
Total Aral Sea Basin	10.023	4.307	1.088	4.045	1.409	0.121	0.067

5. Concluding Discussion

Since the early 1960s, water resource mismanagement has plagued the Aral Sea Basin. The problem assumed a new level of complexity in 1991 when the Soviet Union collapsed and the Aral Sea Basin became a transboundary water resource. Without warning, the development of sustainable and equitable water management practices became the shared responsibility of five sovereign nations each with conflicting needs, goals and priorities. Although collaborating governance bodies were rapidly established, the unsustainable use of water has continued unabated – as have the negative consequences.

Throughout the Soviet era there was one, top down system of governance, responsible for allocating water quotas to irrigated areas and deciding how much was to be harnessed by hydroelectric dams. This was achieved through the Basin Water Organization, a body that remains in existence today (McKinney, 2004). At independence, in the early nineties, the countries of Central Asia began collaborative efforts for managing, using and protecting transboundary water (Djalalov, 2003). They formed the Interstate Water Coordinating Committee (ICWC) in 1992 and the International Fund to Save the Aral Sea in 1993 with the aid of international organizations (UNEP, 1992; Bayarsaiha, 2003; McKinney, 2004). These organizations involved government water ministers and national leaders for the purposes of better managing the water resources, preventing environmental degradation and using water more efficiently (McKinney, 2004).

Progress, however, has been hindered for several reasons, some scientific, others political:

- The water committees focus almost exclusively on surface water in the Syr Darya and Amu Darya catchments and consider only water quantity and not quality (Bayarsaiha, 2003).

- The committees entirely neglect the role of groundwater at the basin level and transboundary groundwater at the international level.

- There is a general lack of interest on a wide enough scale at the political level for proper management. The Asian Development Bank (2003), Djalalov (2003), and McCray (1998) found that the Central Asian organizations were plagued by diverging interests

and sovereignty issues, lacked coordination within management, had a paucity of funds, lacked public participation, and suffered from decrepit infrastructure.
- Even within individual countries there appears to be a lack of coordination on water issues with three different government bodies regulating the three different sources of water (surface water, irrigation water, and groundwater) (Djalalov, 2003).
- Turkmenistan has little interest in working towards Aral Sea initiatives because they are not directly impacted (McCray, 1998).
- The individual economic interests of each country are greater than any benefits from cooperation (Dukhovny, 2003).

Solutions to the region's social, economic, and environmental crises are clearly complex and inevitably long-term. However, none can be successful unless they address the root cause of the problem i.e., water resource management practices that have failed to respect the principles of equity and sustainability. To date, water "management" in Central Asia focuses almost exclusively on the equitable allocation of surface water flows. Such an approach is destined to fail unless the entire water resource including groundwater is considered, and water quality as well quantity is included in the decision-making matrix.

Data suggest that a number of important aquifers occur in the region and many are transboundary. These internationally-shared resources must not only be managed in close collaboration with neighboring countries but also managed conjunctively with surface water as part of a fully integrated water resource management strategy. At present, the importance of groundwater is recognized within individual countries of Central Asia but its regional significance in the context of the Aral Sea Basin is rarely acknowledged.

The importance of managing ground and surface water "conjunctively" within a single integrated management plan has been embraced by the American Water Works Association (1996) in their concept of "Total Water Management" (TWM), which they describe as an exercise in the stewardship of water resources for the greatest good of society and the environment. A basic tenet of TWM is that the supply is renewable, but finite, and should be managed on a sustainable use basis such as to:

- Encourage planning and management on a natural water systems basis through a dynamic process that adapts to changing conditions.
- Balance competing uses of water through efficient allocation that addresses social values, cost effectiveness, and environmental benefits and costs.
- Involve all levels of government and stakeholders in decision-making through a process of coordination and conflict resolution.
- Promote water conservation, reuse, source protection, and supply development to enhance water quality and quantity.
- Foster public health, safety, and community good will.

Importantly, a good water resource management strategy is not limited to quantity but must include programs for water quality protection. In some cases this may include strict control on land uses that may have a negative impact on the water resource. Key components of a water resource management strategy would include:

- Detailed water budget for the watershed.
- The identification and quantification of all significant water withdrawals, including domestic water supply, irrigation, and industry.
- Maps of areas of groundwater vulnerability that include characteristics such as depth to bedrock, depth to water table, the extent of aquifers, and recharge rates.
- The identification of all major point and non-point sources of contaminants in the watershed.
- Numerical models that describe the movement of both ground and surface water in the watershed and how this will respond to climatic change and changes in water withdrawal.
- Numerical models capable of describing the fate of chemicals released in the watershed.
- A comprehensive monitoring program.
- The direct involvement of all stakeholders in the decision-making process including the active participation of representatives from all stakeholder countries where transboundary waters are involved.

In Central Asia, the transboundary nature of the region's water resources makes equitable and sustainable management of the Aral Sea Basin a daunting challenge. Nevertheless, the task is achievable if the problem is approached with sound science and good data, appropriate institutional arrangements, collaborative research and adequate monitoring, with the aim of developing water resource management programs that fully integrate transboundary aquifers as a key component.

References

American Water Works Association Research Foundation (1996). Minutes of Seattle Workshop on Total Water Management, Denver, Colorado, August, 1996.

Bayarsaiha, T (2003) Shared Water Resources: Whose Water? Asian Development Bank. www.adb.org/Documents/Periodicals/ADB_Review/2003/vol35_1/shared_water.asp Accessed 2 August 2006.

Benduhn F, Renard P (2004) A dynamic model of the Aral Sea water and salt balance. Journal of Marine Systems 47:35–50.

BGR (2006) WHYMAP – Groundwater Resources of the World – Transboundary Aquifer Systems 1:50,000,000.

Bjorklund G (1998) The Aral Sea: Water Resources, Use and Misuse. Kiessling KL (ed). In: Alleviating the Consequences of an Ecological Catastrophe. Conference on the Aral Sea – Women, Children, Health and Environment. pp. 42–50.

Central Asian Water. Water Resources of the Aral Sea Basin. Groundwater: Reserves and Uses. www.cawater-info.net/aral/groundwater_e.htm Accessed 18 August 2006.

Djalalov A (2003) Rational Water Resources Uses in Market Economy Conditions. Prepared for the Third World Water Forum, Kyoto, 18 March 2003 Japan.

Dukhovny VA (2003) Integrated Water Resource Management in the Aral Sea Basin. International Committee for Water Coordination (ICWC). Third World Water Forum.

Eckstein Y, Eckstein GE (2005) Transboundary Aquifers: Conceptual Models for Development of International Law. Ground Water 43 (5), 679–691.

Gascoin S, Renard P (2005) Hydrological balance modeling of the southern Aral Sea between 1993 and 2001. Hydrological Sciences Journal 50(6):1119–1135.

Grebenyukov PG (2001) Interaction between Surface and Subsurface Waters: Case Study of a Region in Kazakhstan. Water Resources 28(1):22–28.

Howard K.W.F. 2004. Approaches to the evaluation and protection of groundwater and surface water in situations with competing regional uses. In Teaf, C. et al. (eds.). Risk Assessment as a Tool for Environmental Decision-Making in Central Asia NATO Science Series IV Earth and Environmental Sciences 34. 87–112.

Interstate Water Coordination Commission (ICWC) (2002) Diagnostic Report on Water Resources in Central Asia. Scientific Information Center Interstate Water Coordination Commission of Central Asia.

IWMI (2005) Ground Water Management for Improved Irrigation in the Syr-Darya River Basin in the Ferghana Valley.http://www.iwmi.cgiar.org/centralasia/html/fergana-gwm.html Accessed 10 October 2006.

Jarsjo J, Destouni G (2004) Groundwater Discharge into the Aral Sea after 1960. Journal of Marine Systems 47:109–120.

McCray TR (1998) Enviro-Economic Imperatives and Agricultural Production in Uzbekistan: Modern Responses to Emergent Water Management Problems. Ph D Thesis, University of Kansas.

McKinney DC (2004) Cooperative Management of Transboundary Water Resources in Central Asia. In: Burghart D, Sabonis-Helf T (eds.) In the Tracks of Tamerlane-Central Asia's Path into the 21st century. National Defense University Press.

Morris BL, Litvak RG, Ahmed KM (2002). Urban Groundwater Protection and Management : Lessons from developing cities, Bangladesh and Kyrgyzstan. In Howard, K.W.F. & Israfilov, R. (eds.) 2002. Current problems of hydrogeology in urban areas, urban agglomerates and industrial centres. NATO Science Series IV Earth and Environmental Sciences 8, 77–102.

Morris B, Darling G, Gooddy DC, Litvak RG, Neumann I, Nemaltseva EJ, Poddubnaia I (2005). Assessing the extent of induced leakage to an urban aquifer using environmental factors: Bishkek. Hydrogeology Journal 4:225–243.

Pozdniakov SP, Shestakov VM (1998) Analysis of groundwater discharge with a lumped parameter model, using a case study from Tajikistan. Hydrogeology Journal 6:226–232.

Puri S (ed.) (2001) Internationally Shared (Transboundary) Aquifer Resources. Management – Their significance and sustainable management – A framework document. IHP–VI, IHP Non Serial Publications in Hydrology, UNESCO, Paris. 71 pp.

Puri S., and Aureli A (2005) Transboundary Aquifers: A Global Program to Assess, Evaluate, and Develop Policy. Ground Water . 43, No. 5, 661–668.

Saiko TA, Zonn IS (2000) Irrigation expansion and dynamics of desertification in the Circum-Aral region of Central Asia. Applied Geography 20:p349–367.

Ulmishek GF (2004) Petroleum Geology and Resources of the Amu Dar'ya Basin, Turkmenistan, Uzbekistan, Afghanistan and Iran. US Geological Survey Bulletin 2201-H.

UNEP (1992) Diagnostic Study for the Development of an Action Plan for the Conservation of the Aral Sea. UN.

UNDP, UNEP, OSCE, NATO (2005) Environment and Security: Transforming risks into cooperation. Central Asia: Ferghana/Osh/Khujand area or Ferghana Valley.

Veselov VV, Panichkin V, Zakharova NM, Vinnikova TN, Trushel L (2004) Geoinformation and mathematical model of Eastern Priaralye. Mathematics and Computers in Simulation 67: 317–325.

Yamada C (2005) Third report on shared natural resources: transboundary groundwaters. International Law Commission: Fifty-seventh session A/CN.4/551. Geneva.

INTEGRATED WATER MANAGEMENT

J. STAES
Ecosystem Management Research Group
University of Antwerp, Dep. Of Biology
Universiteitsplein 1, 2610 Wilrijk, Belgium

H. BACKX
Ecosystem Management Research Group
University of Antwerp, Dep. Of Biology
Universiteitsplein 1, 2610 Wilrijk, Belgium

P. MEIRE
Ecosystem Management Research Group
University of Antwerp, Dep. Of Biology &
Chair of Integrated Water Management
University of Antwerp, Institute for Environmental Sciences
(Director of the Integrated Water Management Pilot Studies)
Universiteitsplein 1, 2610 Wilrijk, Belgium

What do we wish to manage?
"Over the past decades, the ecosystem approach has been developed into a strategy, which is now part and parcel of integrated water resources management. Many examples prove that it is often more cost-effective to maintain, or even restore or create, water-related ecosystems than to try to provide the same services through expensive engineering structures, such as dams, embankments or water-treatment facilities."

(Enderlein and Bernardini 2005).

Abstract: The natural water system provides direct or indirectly numerous goods and services. For a long time these goods and services have been used to support society in various and important ways. We can distinguish visible and fast renewable resources as fish, crops, timber, and drinking water that distinctively can be linked to the water system. The water system also supports society through direct or indirect ecological services. The strong interdependence from the water system initially forced a certain harmony between the water system and its users. A combination of growing needs and a technological ability has resulted in an increased control and manipulation of the water system. These developments have also led to a serious degeneration of the system's carrying capacity. The emergence of the first major environmental problems has led to the development of a fragmented water management approach. Initially focused on finding technological "end of pipe" solutions to maintain anthropogenic user functions. Over the last few decades, this fragmented and compartmental approach has shown little adequacy and led to a growing awareness that an integrated and holistic approach is necessary. The term "integrated" has been given many definitions and is still subject to various technical interpretations. Although the integrated water management concepts are intended to be holistic, it has been found that many practical attempts toward a truly integrated approach still focus heavily on controlling the water system to provide specific goods and services at the right time and place (Holling and Meffe 1996; Briggs 2003). Modeling efforts seldom consist of a truly holistic approach, and rather focus on aspects that are predictable, quantifiable and controllable (e.g., hydrology, water quality, etc) leaving out those aspects that are not. Many ecosystem services will never be eligible for command and control because they are involved with a complex web of processes that operate on various scales in time and space (Jakeman and Letcher 2003). An aversion of policy makers toward the unknown, uncertain and uncontrollable makes these services subordinate to those services that are controllable. This is reflected by the aversion of policy-makers to tools and models that try to deal with this complexity (Gustavson, Lonergan et al., 1999; Welp 2001; Quevauviller, Balabanis et al., 2005). Nevertheless, there is enough circumstantial evidence that the decline of vigor and resilience of our natural environment can be ascribed to the rigid management of rivers and its interwoven ecosystems. In this article, we look at concepts of catchment-level ecosystem functioning that should be considered in an integrated water management approach. Finally a conceptual framework for the development of river subbasin management plans is suggested.

Keywords: Integrated water management, ecosystem management, water system theory

1. Problem Definition

Water is a natural resource that cannot be bounded, and the consequences of pressures on water systems can be the cause for conflict. The conflicts and issues that reach policymakers are to a certain extent the consequences of the real underlying problems. The presence of human life and its activities means that we use and thus impact ecosystems. We often adapt and modify our environment to our needs in an attempt to create additional benefits. These seemingly 'incremental' benefits are more often a trade-off with the natural services that ecosystems provide to our society. These natural 'goods and services' are often unaccounted and "invisible". Where ecosystems are overexploited, their ability to provide these goods and services can be lost (Jewitt 2002).

The concept of a catchment as a basic management unit can be considered as an appropriate alternative for the existing administrative boundaries (Odum 1971; Jewitt 2002; Wester, Merrey et al., 2003). A consistent and logical hierarchy of increasingly smaller subcatchments allows "zooming" in and out (Jewitt 2002). Social and economic activities within a watershed have to be consistent with the hydrological and ecological needs of human well-being (IWRA, Agarwal et al., 1999; Lundqvist and Falkenmark 2000). The various impacts of human activities on the generation of ecosystem services must therefore be taken into account. To reach these objectives, there is a growing need for integrated water management plans on international and other jurisdictional levels. River subbasin management plans can play a key part in integrated water management as they can serve as building units for international river basin plans. As there is are growing interactions among science, management and the broader views of the natural environment, policymakers nowadays, rely more and more on science and models to provide decision support (Gibbons, Limoges et al., 1994; Jakeman and Letcher 2003). This shift is driven by a growing awareness of system complexity as history has shown repeatedly that manipulation of natural systems can have unexpected, unwanted and irreversible side effects (Ludwig, Hilborn et al., 1993). Notwithstanding many billions of euro's that have been spent on water management over a very long time span, society is still confronted with major floodings, droughts, and various water quality problems.

The challenges faced in watershed management have reflected the need to deal with spatial variability, scaling and the need to consider explicit linkages among hydrology, geochemistry, environmental biology, meteorology and climatologic variability. But how many of these management plans and models are truly integrated and explicitly consider the interrelations among ecosystem services and ecosystem functioning? If integrated, the growing set of tools for both decision-making and implementation would suggest that the response of policy makers to environmental problems would become significantly improved. Some authors suggest that the solutions to environmental problems do not follow this trend and that there is a gap between the availability of management tools and the actual use (Clark W. and Munn R 1986; Ludwig, Hilborn et al., 1993; Light, Gunderson et al., 1995). This can be explained by the lack of hard scientific arguments for decision-making. But what if these hard scientific arguments are unreachable and inherent to natural systems? Models that are able to integrate these concepts can be of enormous complexity, demand unrealistic spatial and time step resolution monitoring data and need to be run with numerous scenarios as they are driven by unpredictable driving forces such as climatic variability (Pallottino, Sechi et al., 2005). Jakeman (Jakeman and Letcher 2003) state that integrated assessment modeling has failed because it was typically unable to deal with the many connections and complexities. Born and Sonogni (Born and Sonzogni 1995) conclude that this inability has led to reductionism, focusing only on a portion of the watershed and losing touch with the larger picture. On the other hand there is no time to wait for adequate integrated assessment that complies with the conditions of Parker (Parker, Letcher et al., 2002). Climate variability, model process complexity, scenario exploration, long term leads and lags, lack of monitoring data that addresses the spatial and temporal variation, software incompatibility etc. make the aspiration for accurate, quantitative, numerical and integrated models unrealistic in the short term.

The conceptual point of modeling however reflects perfectly the needs that are put forward by our "command and control" society. As we are entering an era in which it becomes clear that ecosystem services are under such pressures that they might collapse, there is a relative sudden necessity and political interest to take action. Ironically, at the same time we still have a tendency to reach back to the same active management and optimization we applied to overexploit the ecosystem goods and services. We seem to have forgotten that natural systems have self-organizing and regulating capabilities. In order to provide a fast and adequate answer, we need to have more control on the impacts on the one hand and drop control on ecosystem functioning on the other hand. We need to look at the fundamental building blocks of large scale system functioning and balance them according to the needs within a basin perspective. Environmental issues are often the result of numerous unknown and complex interactions. Likewise as our socioeconomic environment is highly unpredictable, so is the environment. This fact is often difficult to accept by policy makers and engineering scientists. Finding hard and conclusive answers with guaranteed results can be difficult when dealing with natural and open systems. For our society, which is dominated by short-term command and control management, environmental problems remain largely unanswered in the long run. This pathology of command and control (Holling and Meffe 1996) is still relevant today (Briggs 2003) and still leads to further loss of resilience. The risk aversive and passive attitude toward environmental problems is often embedded within policy institutions. The reasons lie partly in the ambivalent nature of the choices to be made. The complex nature of ecosystem functioning and the associated uncertainty make it harder for policy makers to take risky decisions (Hosseini 2001). Environmental management failure is partly due to a mismatching among spatial and temporal scales of resource management decision-making, and the scales at which the ecosystem processes operate (Levin 1992). Policy makers tend to delay dealing with problems, then try quick fixes resulting in "crisis management". Often a whole stack of natural resource management programs, studies and reports are conducted without adequate implementation or evaluation of their effectiveness. There is inadequate interaction among the various policy domains, leading to multiple, unlinked and incompatible plans and planning processes (Briggs 2003). The term "integrated" is often used to describe many forms of integration (Downs, Gregory et al., 1991). There are at least five different types of integration that can be identified as integrated assessment modeling (IAM). Truly integrated assessment functionally connects various stakeholders, scales, models and disciplines for the consideration of integrated management issues (Parker, Letcher et al., 2002). It is however acknowledged by many model developers that although "integrated assessment and modeling" is often strived for, it is never truly achieved and, perhaps, even impossible to achieve (Parker, Letcher et al., 2002). In practice, there is little attention given to the shortcomings of applied IAM. So far, various users equated the efficiency of water management with maximum use of water resources. Environmental and ecological considerations as well as downstream users were given little attention.

Determinism in the light of positive sciences is the idea that everything can, in principle, be explained. Applying determinism to ecosystem functioning would mean that events could be explained or predicted by combining input variables (antecedent events) into functional relationships that follow the laws of nature. Determinism is likely the only way to conceptualize the ecosystem functioning in all its complexity. It will however not be reached in all its aspects as the input variables will unavoidably show non-quantifiable and highly variable statistical properties. If the relationships were exactly known, we would be able to express the risk statistically as a rational approach toward variability. The functional relationships are however based on theories, are not exactly known and in addition they inevitably need to be calibrated with input data that reflect the same or similar statistical imperfections. Thus even risk will always reflect the intrinsic properties of statistical uncertainty. The statistical properties of the input variables also apply to the output variables in terms of calibration. In many cases, the range of the output variables is known, but not their statistical properties. We use scenarios and models to calculate a range of expected outcomes, but are unable to define probabilities. If crucial functional relationships are based on assumptions, we need to be very cautious in how these assumptions are applied. Indeterminacy is easily reached if functional relationships are too complex or numerous, or if the monitoring of input variables becomes unachievable by the sheer number of variables or the needed spatio-temporal resolution. This has also been expressed by others: Determinism, Risk, Uncertainty, Ignorance, Indeterminacy (Wynne 1992; Harremoes and Madsen 1999; Harremoës 2000; Harremoës 2003; Walker, Harremoes et al., 2003; Walker and Marchau 2003).

Simonovic (Simonovic 2000) addresses the future of water management tools and clearly describes two important paradigms (i.e., the complexity paradigm and the uncertainty paradigm) regarding model-development, but omits the importance of ecosystem behavior and functioning. These two paradigms are therefore projected onto the modeling of water resources management/allocation but not integrated ecosystem management. The complexity paradigm states that decision-making will be more difficult due to increasing domain complexity. The numbers of aspects that have to be considered are numerous in a densely populated system, with numerous stakeholders and an already heavily modified environment. The temporal and spatial scale that need to be considered increase as transfers of human activities, substances and water between basins (or other management units) become more important. As the socioeconomic system increases in complexity, there is a need for more agencies, water managers that operate their competences at different policy levels and geographic units. The uncertainty paradigm implies an increase of uncertainty in both the time domain as well as in the spatial domain as we progress to the modeling of more complex and spatially distributed processes. The inherent variability and heterogeneity of our environment can puzzle not only the developers of models, but also its users. A fundamental lack of knowledge regarding the functioning of the water system in all its aspects causes uncertainty by making assumptions and simplifications of the real world (Simonovic 2000). But one resulting problem with such 'realistic' assumptions and simplifications may be a lack of sensitivity in the hydrological models in representing what the impacts of the scenarios will be (Jessel and Jacobs 2005). But even if a model can capture the complex interrelations, there is in many cases a fundamental lack of data and information (Radwan, Willems et al. 2003), as the same variability and heterogeneity demands intensive monitoring involving the following activities:

- Increasing complexity and a larger number of variables.
- Higher temporal resolution (smaller model time-step).
- Higher spatial resolution.

We consequently need to sample more variables, more frequently and at an increasing number of locations. The design of monitoring strategies also becomes more important (Schroder, Schmidt et al., 2003). Nevertheless we need to safeguard that the monitoring itself does not become more expensive than the goal it serves. We should of course distinguish the various reasons for modeling. But if we consider that the model serves to provide output for active management and control, the condition of cost and benefit of additional monitoring should be fulfilled. The benefits of increased supply of certain goods and services should, leaving the costs of ecosystem functioning losses out of consideration, at least be compared with the costs of monitoring and management infrastructure.

Although we have a growing knowledge regarding system analysis (Wurbs 2005) and an increasing computational power to process and analyse information, it depends whether this evolution will outrun the increasing complexity. A crucial factor will be the availability and quality of information we can apply. At appropriate scales, it is however possible to overcome this variability and heterogeneity and find a general behavior of system components. Generalised and large-scale indicators can then serve as inputs to assess system behavior. The quest for the right indicator parameters that are inexpensive to measure and easy to understand is important but there is little scientific consensus to be found in terms of their selection (Gustavson, Lonergan et al., 1999).

It can also be questioned whether the relationships found in natural conditions can still be properly represented in a system that is often dominated by urban infrastructures. There is a growing body of research in urban hydrology, drainage, irrigation etc. These systems are interrelated with the rural environment, but act quiet differently (pumps, sluices, piping, basins, and overflow valves), resulting generally in discontinuities over time and space (Beck 2005). As the rural-urban interface increases, we need to integrate two very different systems and deal with the "grey zone" (in between) to which none of the models actually apply. This problem increases when the management of this urban hydrology cannot be captured by management protocols and operational decision support mechanisms. In the last decade, the acceptance and introduction of new management concepts in urban water management has evolved (Harremoës 2003; Beck

2005). For newly designed systems, operational management is often incorporated (Vanrolleghem, Benedetti et al., 2005). The networks of older structures however date from different periods in the urban development, are often badly maintained and little is often known about their behavior.

It has been commented on by several authors that most models fail at the rural-urban interface. The time and scale of the urban water system is also very different from the natural environment. Urban systems comprise man-made infrastructures; pumps, pipes, treatment plants etc. and these elements have aspects that are more easily controllable because they are locked within the system. This is of course only the case if the system and its functions are identified as completely as possible. It is easier to model transport through pipes and pumps than transport through river networks and groundwater flows. It is also easier to create control points and devices within an urban system as they have a centralized character (non-point sources and distributed transport mechanisms are hard to control). As a result, the time and spatial scale of an urban system is much smaller that that of a watershed as it is averaged out over hydrological events within a long-time perspective (e.g., nutrient leaching will be distributed over various rainfall events). The urban control is often expressed in hours and meters while the watershed management is expressed in units of months and hectares. This higher level of control can be of major importance within certain basins because there is real control over the output to the river network that can be managed with classical engineering control methods. The separated rain and wastewater drainage system are examples of new technology that is intended to increase control.

Because of these advantages, there have been attempts to harness the natural system using engineering projects. These attempts have been generally unsuccessful, because they do not consider the functions provided by the natural systems. On the other hand we can only hope that the urban islands within the natural systems can be controlled as far as possible because they have lost most of their regulating capacity and thus provide little or no environmental functionality. Integrating these urban water systems within a watershed system is a necessity. There are however some issues: the urban systems are not always that much an island as we wish them to be. Some areas are so fragmented that there is a gray zone between the urban and the natural system. The man-made infrastructure can be seen as discontinuities in an environment that is shaped by natural processes. Although there is enormous variability within the natural environment, patterns exist at larger scales. Additionally they can be linked to abiotic factors such as morphology. Man-made infrastructures cut through these patterns and interaction takes place. Water and substances are transferred and transformed in ways that are opposite to the natural pathways that exist. Secondly existing infrastructure and its functioning is often unknown due to the replacement (e.g., adding of new infrastructure to the old sewage system). As a conclusion it can be said that it is ambiguous to assess the effectiveness of efforts for integrated watershed management, as there are two sides to the effectiveness, namely the control level and the magnitude level. As the control-level rises, the net effect in avoiding critical pollution levels at a specific moment might be more important than an overall reduction of pollution sources that are subjected to climatological influences. The optimum will be somewhere in the middle but the control level should be considered in the cost-benefit analysis of investment evaluations. A thinking exercise about control parameters for urban and watershed systems could be very interesting. These aspects of a control parameter should be characterized in order to assess effectiveness. Some of the aspects can be considered as decision variables to implement a control parameter.

It can be questioned whether a "ladder of integration" should be developed. A conditional stepwise approach to integrated assessment and modeling (IAM) would be desirable as it directly affects the quality of the decisions. Increasing the level of integration would therefore consequently increase the level of uncertainty by an exponential factor. An important methodological problem of IAM is the scale, and the resolution of different system components (Parker, Letcher et al., 2002). Indeed we should be cautious in creating IAM-models that are not transparent, since our ability to model our environment in multiple aspects and scales remains limited. Additionally, socioeconomic modeling within environmental models provides at most a sense of direction regarding the importance or probable consequences of certain decisions. Parker et al. (2002) state, "The issues are the center point of integration as IAM seeks to avoid the fragmented approach". But can we truly define the "issues" in general terms? In this initial step of defining issues, subjective formulation of issues, viewpoints and objectives already compromise the IAM. The first problem is that when trying to implement a Decision Support System (DSS) for basin management, complexity leads to ill-defined and ill-structured problem definition (Reitsma and Carron 1997). There are conditions which have to be met in order to implement a DSS: a) a limited set of objectives, alternatives or control variables, b) a solid physical model that can calculate all state variables for the objectives, and c) the objective equations are numerically and standardized. Note that if it is not possible to find an equation that values the objective, this objective cannot be included. Models are therefore biased toward aspects that are predictable, quantifiable and controllable, excluding e.g., ecological objectives.

The characteristics of both human nature and ecosystem behavior also create circular reasoning patterns in IAM. The explicit and recognized valuing that needs to be applied in many DSS and IAM is subject to feedback loops of supply and demand. The equations that value the objectives are to a large extent subjective and only reflect the value within a certain scale and time frame. The decision outcome might influence the generation of and demand for ecosystem goods and services, which influences in their turn the value of it. Both the socioeconomic world and the environment are controlled by these feedback loops that operate on different spatial and temporal scales. IAM are often built on a series of subjective and dogmatic rules and choices. Which are the objectives to include? Which variables are considered for those objectives? What are the spatial scale and the time frame for assessment? Secondly it is stated by Parker et al., (2002) that the different values represented by multiple stakeholders groups need to be recognized and explicitly included in models. An important issue here is that the value for future generations (the option values) is typically undefined. Scenarios for the future are uncertain, with possibly different stakeholder groups using different ecosystem goods and services. The aspect of reversibility and how much we can alter in an irreversible state, without compromising them is crucial. Secondly, the ecosystem itself is not represented and neither stakeholders nor government are able to define objectives.

Synonyms like "integrated management, sustainability, ecological integrity" are easily utilized by policy makers but are hard to quantify and can be viewed from different perspectives and at different time horizons. Industrialized society is founded upon historic assumptions that generally externalize the benefits gained from catchment functions (Everard, 2004). Placing values on decisions that influence system functioning is difficult because societal appreciation is highly susceptible to externalities, resulting in unpredictable scenarios over longer terms. The positive and negative values of a decision need to be considered over the complete time frame of its influence. Expressing the environment as an economic value holds risks, as economical valuation can be determined by the rules of supply and demand which is strongly influenced by societal purchasing power. In other terms it means that the appreciation of ecosystems and nature is much higher when you have a well-developed society (education, employment, and social welfare). Secondly ecosystems generally do not reflect the 'market value' of its appreciation (Scheffer, Brock et al. 2000). This is due to the irreversibility of ecosystem losses and the time lag between environmental impact and its associated consequences. Re-creation and even restoration of high quality nature is extremely difficult and many attempts have failed to meet the objectives (Hobbs and Harris 2001; Klotzli and Grootjans 2001). The reference point of the environmental state also shifts away with time (generations) making the appreciation value of further ecosystem losses non-cumulative. The value of an ecosystem is also susceptible to the option-value theorem as ecosystems provide a potential to generate a whole range of ecosystem goods and services (De Groot, Wilson et al., 2002). The option of delaying or waiting to decide has been found to be a valuable and often overlooked source of value in a project and for many decision support models. This concept originates from economic engineering and is applicable to any universal uncertainty decision-making. Herath (Herath, 2001) gives an overview on the option value concept and the decision models which are based on this concept. Examples of environmental oriented applications include: forestry (Insley, 2002), land-use planning (Schatzki, 2003), and biodiversity preservation (Kassar and Lasserre 2004). The Option Rule is that one should invest today only if the net present value is high enough to compensate for giving up the value of the option to wait. Because the option to invest loses its value when the investment is irreversibly made, this loss is an opportunity cost of investing. In adopting environmental policy, there are two types of irreversibility (Pindyck, 2002). – policy irreversibility and environmental damage irreversibility. They both originate from uncertainty and irreversibility. The choice to implement or not imposes a chance for costly scenarios when not making the right choice. The first scenario is when implementation of a policy has been done, but it wasn't necessary and the second scenario is when there was no policy effectuated (or too late) and irreversible damage has been done to the environment. The goods and services that are provided by ecosystem functioning has a similar option-value, as the loss of ecosystem functions is mostly irreversible in the short term.

The concept of ecosystem services has become firmly established in the vocabulary of ecological economics. It also seems to have gained attention in wider scientific circles and among some policy-makers. This is at least partly likely to be due to publications such as Hansson (Hansson, Bronmark et al., 2005), De Groot (De Groot, Wilson et al., 2002; De Groot, 2005), Jewitt (Jewitt, 2002), Baskin (Baskin, 1997), Costanza (Costanza, d'Arge et al., 1997) and Daily (Daily, 1997). Although the socioeconomic importance of ecosystem goods and services has been increasingly acknowledged by policy makers, it is still a concept that raises many questions since it cannot always be quantified economically. The definition of "goods and services" implies that we have to estimate a "potential value" that would be lost when we lose the potential of generating goods and services.

Secondly there are issues regarding the scale that we observe. In order to asses these ecosystem functions and their natural 'goods and services' it is important to look at different scales. A watershed ecosystem can be seen as a complex web of ecosystems that are interlinked by longitudinal, lateral, vertical and temporal dynamics. Structural and dynamical complexity or even chaos can be seen as a condition for stability of ecosystems and the services they provide (Upadhyay and Rai 1997). Nevertheless there are driving economic forces that result in natural systems being manipulated and altered for the exploitation of certain ecosystem goods and services (De Groot, 2005). These manipulations can damage the carrying capacity and stability of the overall system endangering other ecosystem functions in the long run.

The carrying capacity (CC) of a river basin is its ability to sustain a certain amount of human activities. The developments by humans for many thousand years can be characterized by continuously trying to increase the CC. To a certain extent, the manipulation of the water system can be acceptable and indispensable to support a stable socioeconomic system. Often "improvement of CC" is only temporary, benefiting only specific user groups or having dramatic effects on other aspects of the water system. Most of the basin management plans, models, participation, decisions support systems focus mainly to control those ecosystem goods and services that have been acknowledged to have current socioeconomic importance. This approach of management leads inevitably to failure as other "goods and services" might disappear, primarily due to the use of a sectoral and segmented approach, and the following consequences may ensue:

- Using water supply management instead of water demand management.
- Emphasizing flood defense instead of upstream retention.
- Focusing on drainage and irrigation instead of water management based on land-use planning.

Over the years, a great deal of research has focused on the development of ecosystem functioning concepts. Many concepts have been proven to exist and have been acknowledged, but only a few have been quantified. Complex concepts in water system functioning are often excluded from the decision-making arguments because they cannot be quantified, or are otherwise difficult to measure. Some of these processes are event driven and thus highly unpredictable and/or have high spatial and temporal variation. The complex web of processes involved and the distributed interactions between the different compartments and/or the river-valley system make these concepts hard to quantify in terms of their spatially distributed and multiple effects. Hence their importance is often recognized in principle, but not taken into account. Finally, these processes are still not acknowledged by the mainstream as being valid arguments in making decisions. As these processes require a natural and dynamic water system, they subordinate other uses of the land and suffer considerable opposition as they are often considered to serve the "nature development lobby". The problem of being unpredictable and hard to quantify, makes scientific evidence of their importance hard to substantiate and easy to refute. Although the problems have been studied extensively and sufficient conceptual models have been developed, there are no real answers on how to deal with these complex issues.

Accuracy and transparency are found to be key aspects that make or fail any decision process. Sadly, these two terms do not compare with complexity. As a result, there is perhaps a need to reduce emphasis on developing more complex models and try to focus on working and applying concepts related to their scale utilizing available data. If accuracy is limited by complexity and available data, the use of sophisticated models is useless. Especially in those cases it is of utmost importance to work on the transparency-aspect. A low transparency generates mistrust among partners as it is perceived as a 'magic box' of which they have no understanding, while an open and transparent approach opens doors to discussion and participation and often leads to better results.

From a conceptual point of view it can be stated that certain "hotspots" of the ecosystem (chemical and hydrological processes) are important to sustain the system's overall function (McClain, Boyer et al., 2003). If we are able to pinpoint locations where these intense processes take place, we should value them according to their downstream benefits and upstream pressures. Some of these hotspots can be pin-pointed rather easily as they depend largely on abiotic conditions (e.g., landscape depressions, seepage areas, moorlands, and frequently inundated areas). Protecting these areas is not sufficient as their ecosystem function (ability to generate a range of goods and services) is also dependent on larger scale processes and interactions that need to be preserved.

What if we are proceeding toward a dead end in environmental planning? Should we develop those complex models? Can a far simpler approach not be as effective? Do we continue to outrun ourselves in the myth of our technical ability of control and command? If we are cautious, we use our efforts in minimizing and mitigating interactions between human activities and the environment in order to address:

- The closing of substance cycles (recycling of resources and more efficient use).
- Reducing the contact surface between urban – rural and natural systems (spatial planning).
- Reducing and controlling the transfer pathways between the different systems (technology).

The focus is therefore not on the issues, but on the management of interactions. If we apply such principles such as caution and source-oriented measures, we need to research how interactions take place between anthropogenic activities and the water system on a local scale. This does not mean that we discard water system functioning knowledge, as it is crucial to determine which interactions are important to manage. If we continue to apply the human-oriented management and control to our entire environment, we indeed have to know the functioning of the entire ecosystem.

2. Concepts Used in Water System Theory

Integrated Catchment Management (ICM) is an application of the "sustainable development" concept on aquatic ecosystems and is rather more a way of thinking than an exact science. The term itself was first used by Gardiner (Gardiner, 1984) and has been adapted and supported by many others over the last thirty years. ICM is also adapted to meet requirements and needs of policies such as the European Water Framework Directive (Wasson, Tusseau-Vuillemin et al., 2003). But ICM has been given different interpretations by both scientists as policy makers and has been translated into different definitions (Downs, Gregory et al., 1991).

A first difficulty in the terminology is the mixing of "Integrated Water Resources Management" and "Integrated Catchment Management". The first term is often used in arid regions where the integration is limited to the integration of activities to improve efficiency and/or availability of water. The term integrated is sometimes left behind in book titles such as Water Resources Management by Grigg (Grigg, 1996), perhaps acknowledging that true integration may be unachievable. Is the focus on managing the exploitation of goods and services in order to sustain life support, food-production and industrial development or do we manage our ecological footprint in order to sustain the generation of ecosystem goods and services? In many cases, the focus is rather on supply management than on demand management. In water resources planning, it is often found that other goods and services beside freshwater are excluded. The focus is mainly on increasing water supply by storage and release.

The definition that was given for the Flemish Water Policy Plan (VIWC 2000): "integrated water management is the coordinated and integrated development, management and restoration of the water system, so it can fulfill the quality conditions to the ecosystem and its multifunctional use, without compromising this multifunctionality for future generations. The goals and the inherent choices regarding functions and land-use need to be based on knowledge of water system functioning and its boundaries of carrying capacity." The term "ecosystem management" suffers the same semantic difficulties as sustainable development, stability, carrying capacity, diversity and integrated resources management ((Agee and Johnson 1988), 1988) (Grumbine 1994; Grumbine 1997). What unites these terms is that they refer to systems and their behavior. Systems are not always that well understood in term of their behavior. Much of the rivers in the North American and European temperate zones that have been studied by the early stream ecologists have been significantly altered by human activities and engineering before any research took place (Petts, 1989) cited by (Ward, Tockner et al., 2002). The study of these modified ecosystems led to an overestimation of their deterministic, static, homogenous and scale-independent behavior. A resilient ecosystem is one that has the self-regulating capability to deal with disturbances (both natural and manmade). Ecosystems are shaped by powerful adaptive and self-organizing forces (Levin, 1998). Stability and resilience on a basin scale are a result of numerous interactions of ecosystems processes on different scales (Levin, 1992). Costanza and many others have put forward many thoughts and concepts of "ecosystem health", "sustainability" and "ecological integrity" (Rapport 1992; Hobbs, Saunders et al. 1993; Baker 1995; Callicot and Mumford 1997; Costanza and Mageau 1999). The definition of Costanza is that a healthy ecosystem should be able to maintain structure (organization) and function (vigor) over time in the face of stress (resilience) (Costanza, 1999).

Much work has been done on the study of ecosystems as non-equilibrium systems; that is, with a dynamically stabilized set of ecosystem domains whose structure is maintained by variability (Holling 1992). C.S Holling (1992) has attempted to describe an ecosystem with help of a theoretical model. His theory is that an ecosystem never finds itself in a stable position. The main properties of ecosystems, as described by Holling in 1978, are their dynamic behavior, spatial and temporal heterogeneity, multiple levels of stable behavior and their internal complexity. These systems usually comprise a complex web of subsystems. The extensive stack of ecosystem terminology that exists, points at the inherent complexity of the natural environment e.g., multiple scales, hierarchy, self-organization, variability, diversity, heterogeneity, redundancy, non-linear behavior, hysteresis, feedback, limitation, autocatalytic processes, time lags, accumulation, retention, breakdown thresholds, competition, synergy, interdependence, etc. (Ulanowicz 1986; Holling 1992; Meffe and Carroll 1997; Levin, 1998; Peterson, Allen et al., 1998; Carpenter, Brock et al., 1999; Costanza and Mageau 1999; Peterson 2000; Scheffer 2001; Scheffer and Carpenter 2003).

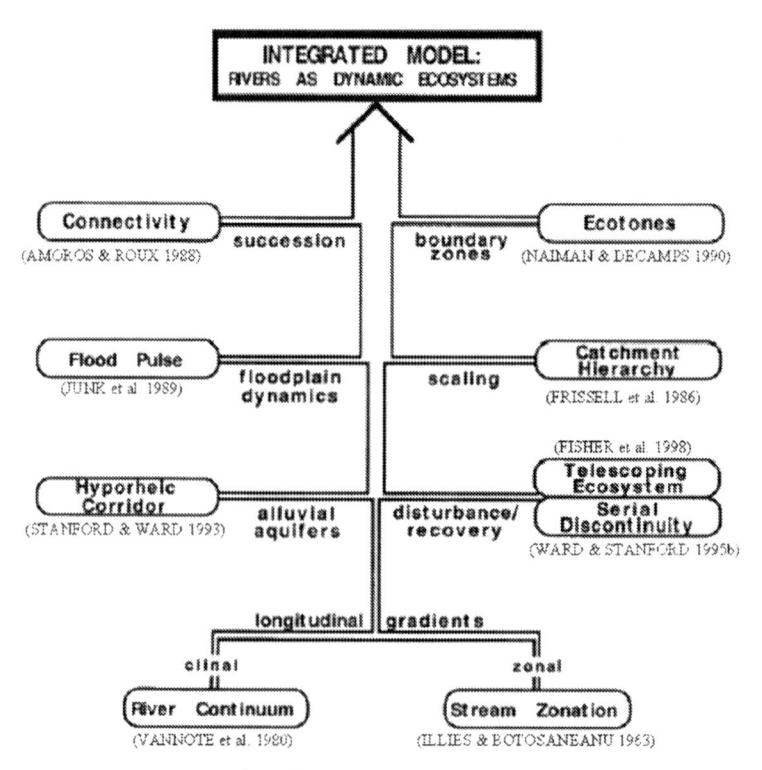

Figure 1. Integration of ecosystem concepts (Ward et al., 2002)

Landscape ecology and riverine ecosystem concepts attempt to characterize the functional units of macro-scale systems. The scheme of Fig. 1 (Ward, 2002) integrates the most important ecosystem concepts. Together, these concepts cover the four dimensions of riverine ecosystem functioning including longitudinal, lateral, vertical and temporal dynamics (Petersen, Petersen et al., 1990; Boon 1992;

Ward and Wiens 2001; Tockner, Ward et al., 2002). The river continuum concept (Vannote, Minshall et al., 1980) is an ecological classification of watersheds on structural and functional characteristics. Strahler (Strahler, 1957), had previously classified rivers on their stream-order. The stream zonation concept (Illies and Botosaneanu 1963) classifies distinct zones along rivers that are separated by transition zones. A common fact is that they consider longitudinal and lateral gradients. The "spiraling concept" (Walker and Marchau 2003) complements that lotical systems tend to have a retention capacity by spiraling nutrients, sediments in the river system. The "flood pulse concept" (Junk, 2003) points to the interaction between the river and its floodplain and links itself with the spiraling concept (Elwood, Newbold et al., 1983; Pinay, Decamps et al., 1999). The "Serial discontinuity concept" (Ward and Stanford 1983; Ward and Stanford 1995; Ward, Tockner et al., 1999) focuses on the impact of dams on longitudinal disruptions in resource gradients. The telescoping ecosystem model (TEM) of Fisher (Fisher, Grimm et al., 1998) specifically addressed the differential recovery trajectories within four subsystems (surface stream, riparian, hyporheic, and parafluvial) following disturbance (e.g., flood, drought). The transformation and exchange of substances along these hydrological interconnected subsystems determine the retention and recovery characteristics of the river system. The hyporheic corridor concept (Stanford Ja 1993) reflects the importance of lateral and vertical dynamics of the multi-channel stream-floodplain interactions. The research concerning aspects of scaling and hierarchy shows that ecosystem processes and functions operating at different scales form a nested, interdependent system, where one level influences other levels above and below it (Allen and Starr 1982; O'Neil, Johnson et al., 1989). Frissel et al., (Frissell C.A and Hurley 1986) classified subsystems (watershed, segment, reach-systems, pool-riffle systems, and microhabitats) on physical characteristics and their spatial and temporal characteristics. The approach of Petts & Amoros (Petts and Amoros 1996) is similar, though it focuses more on the aspects of functional units. Both approaches unite themselves in the fact that they define their systems on their forging processes and their specific physical and biological structures. Connectivity-theory originates from migratory and gene flows between populations (Merriam, 1984), but can be applied to water systems. The hydrological connectivity is a pathway for the transport of energy and matter between subsystems (e.g., groundwater-surface water, river-floodplain) and consequently strongly influences the succession trajectory and the biotic communities (Amoros, Roux et al., 1987; Tockner K and Ward 1998) of those subsystems. In addition, there are many functional processes that are driven by hydrological connectivity (Ward, 1998) e.g., denitrification along aquatic-riparian gradients. Ecotones can be defined as transition zones between relative homogenous patches. These zones have a relative richness in processes and biodiversity when compared to their surroundings (Naiman and Decamps 1990; Zalewski, Puchalski et al., 1994; Ward and Wiens 2001; Zalewski, Bis et al., 2001). Mostly they have a wide range of natural disturbances and dynamics leading to spatial and temporal diversity. Their role in energy and material fluxes has been probably underestimated as their role is often suppressed as represented in managed river systems. Further, conceptual approaches have dealt with the production and processing of organic matter, such as the nutrient spiraling concept (Newbold, Elwood et al., 1981; Elwood, Newbold et al., 1983).

How do we translate these interrelated concepts into practice? In ecosystem management there is a strong focus on maintaining a certain type and level of ecosystem services, while focusing on the generation of ecosystem services. Resilience and stability can be achieved by allowing certain natural processes to occur, as change is a natural resultant of their dynamic behavior and thus a part of the resilience. Freezing ecosystems in a certain stage leads to failure (Pavlikakis and Tsihrintzis 2000).

The objectives and arguments of nature conservation and the new theorems on ecosystem health differ on some aspects and this seemingly implausible conflict is still a subject of debate (Pickett Sta 1992; Talbot 1996; Fiedler Pl 1997; Pickett Sta 1997; Yaffee 1999). Conservation ecologists like Goldstein (1999) criticize the ecosystem approaches for being general and vague and therefore not useful to conservation management as they are unable to set boundaries, prioritization or management objectives. The vagueness and circularity of the ecosystem management terminology is considered a threat to conservation and biodiversity. The compelling complexity has led to a strategy of generalization in order to seek answers. The ecosystem management concepts and related terminology does not exclude anything and are not helpful in simplifying or clarifying management, but rather serve to confuse and obscure (Goldstein, 1999). Nature conservation aims primarily for the protection of biodiversity and endangered species and consequently tries to limit disturbances that might destabilize the ecosystems in which those species reside. Allowing disturbance and dynamic behavior will lead to irrecoverable biodiversity loss if the redundancy of those systems has become scarce (diversity, scale, and connectivity requirements). Consequently traditional conservation ecology can be seen as a necessity because of massive ecosystem loss. In that respect both approaches are justifiable and necessary. In the long run there are difficulties to maintain these island ecosystems within a changing environment. To provide sustainable biodiversity, nature conservation should also focus on macro-scale diversity in terms of restoring ecosystem diversity and connectivity.

A conclusion might be that we should recognize that the complex, adaptive nature of ecosystems (Levin, 1999). To build in environmental security, we need to preserve a level of unconfined ecosystem functioning and learn from their behavior and the services they provide.

There are reoccurring management principles that can be applied to maintain system resilience on any scale (Yaffee 1999). Grumbine (Grumbine 1994; Grumbine 1997) presented ten themes that need to be considered for environmental management: (1) Hierarchical context (2) Ecological boundaries (3) Ecological integrity (4) Data collection (5) Monitoring (6) Adaptive management (7) Interagency cooperation (8) Organizational change (9) Humans embedded in ecosystem (10) Values. Carpenter and Levin (Carpenter, Brock et al., 1999; Levin 1999) (1999) formulated the following eight principles that apply specifically to the attitude that one should have for adaptive management.

1. Reduce uncertainty
2. Expect surprise
3. Maintain heterogeneity
4. Sustain modularity
5. Preserve redundancy
6. Tighten feedback loops
7. Build trust
8. Do unto others as you would have them do unto you

Basin management and land-use planning should be adaptive and allow dynamics and heterogeneity to a certain extent. Therefore certain areas should be designated to allow self-organization. Building resilience means creating a diverse and spatial heterogeneity in ecosystems, structures and in which we allow processes to take place with minimum control and command. Balancing ecosystem diversity and functionality in a basin perspective translates itself in minimum requirements for certain processes to occur.

Integrated water management software solutions have been developed and are easier to assess as they are easier to compare and are well-described in accessible literature. Existing integrated models appear to be poor on the ecological side and few of them have incorporated anything beyond water quality models. Biological community dynamics models are rarely linked to existing catchment models. Furthermore, these integrated systems lack the ability to exchange/swap models of particular domains, in order to improve efficiency, effectiveness of modeling particular processes or, for example, to use existing (legacy) models for a particular problem (Hutchings, Struve et al., 2002). The integration of these models is confined by incompatibilities of scale, dimensionality, parameter meaning, process control, and modeling platforms. However, the integrated viewpoints of the different disciplines normally encountered in the development of a master plan would not converge on the common objective of reaching sustainability unless a clear unified methodological approach is devised to value the different objectives – often conflicting- encountered in the planning exercise at a basin level. A cost-benefit analysis seems to be a way to establish the relative importance of the different issues involved but has the drawback of having to assign cost and benefits to environmental effects and assets (Hutchings, Struve et al., 2002).

3. An Approach for Adaptive Management in Flanders

Land-use in Flanders is intense and characterized by high fragmentation (Antrop and Van Eetvelde 2000). The total developed area represents more than 14% of the total territory, while 62.4% is occupied by agriculture. Industrial areas are dispersed. There has been little attention for spatial planning that reckons with land-use suitability (soil quality, hydrology, etc.). The resulting fragmented land-use has placed a mortgage on future developments of any kind and has had a significant impact on hydrology and ecological functioning (Antrop and Van Eetvelde 2000; Gulinck and Wagendorp 2002). Many problems in water management originate from unfortunate choices in spatial planning and land-use practices. Land-use patterns that reckon with the physical properties of soil and hydrology cause less interaction with the water system whilst a high discrepancy between actual land-use and physical suitability urges a more intense adaptation of the system and thus to a higher impact of land-use on the water system. An increased technological ability to alter the natural system has contributed to the detachment of land-use development from the physical system. A strong interdependence from the water system initially forced a certain harmony between the water system and its users. Many of the valuable cultural landscapes still witness the traditional rural landscape of before the important changes since the 18th century caused by the industrial revolution (Van Eetvelde and Antrop 2005). The basin management plans need to address a long-term vision of spatial planning that reckons with land-use suitability. Often there is a lack of integrated management, as well for water management as for the other policy issues. The interaction between socioeconomic activities and the environment is obvious. This results not only in deteriorating environmental quality, but poses problems for the socioeconomic functions provided by the environment. To narrow it down to water management it is seen that there are conflicts between upstream and downstream functions. As pressures on land-use and water resources rise, they do not only interfere with each other on a local level, but there is also a longitudinal conflict. To provide a long-term sustainable use of the water system, there is a need to define limits to protect the carrying capacity. In order to reach sustainability there has to be a balance at any point between the upstream supply of goods and services and the downstream use of goods and services. Any imbalance results in an unsustainable situation as the downstream activities and functions are endangered. To balance supply and demands there is a second tension field between the ecological and societal use of system resources. Often it is neglected that these systems need specific conditions to maintain their functioning (e.g., water quality, quantity and dynamics). A history of command and control over the water system might in some cases lead to a deadlock. As pressures on land-use and water resources rise, they often interfere with each other and often the least demanding player survives the game. There is no need to organize the use of land and water until the pressures start to cause effects in changing the water system characteristics. At that moment, the land-use and the demand for water resources have often been diversified making it impossible to satisfy all. This diversity requires that choices have to be made in order to create a sustainable situation. Because this socioeconomical situation is historically grown, it is politically very difficult to impose strong measures. The implementation of non-diversified measures on the other hand will lead to the same effect, as the margins for the different stakeholders might be also very different. In combination with a complex and often highly anthropogenic-influenced water system, water management has become a difficult task. The application of theoretical ecology to river conservation is one of the first out of ten recommendations according to (Boon, 1992). It has been argued that practical river restoration measures should allow free system functioning to a certain extent and within certain geographical boundaries (Petersen, Petersen et al., 1990). The physical system needs to be considered as a structuring factor in land-use planning. Social and economic activities in the drainage basin have to be consistent with the hydrological and ecological needs for human well-being (Allan, Erickson et al. 1997; Lundqvist and Falkenmark 2000). Several concepts have been established in respect to land-use patterns and functionality such as Land Quality (Bouma 2002; Bouma 2006), Leakiness Indices (Karlen, Mausbach et al., 1997; Doran and Zeiss 2000; Dumanski and Pieri 2000) and the Dissipative Ecological Unit (Ripl 1995; Ripl 2003). Shared by these approaches is that fluxes of water and substances determine the sustainability at certain locations given a certain land-use. Management should therefore delineate three types of function priorities on river basin level:

1. Areas in which anthropogenic functions dominate. The management challenge is to manage system behavior to provide the needed goods and services. Secondly, there should be measures taken to minimize and control the impacts of the activities.

2. Multifunctional areas. In these areas there should be an interaction between the exploitation of goods and services and the water system/ecosystem functioning. Multiple anthropogenic functions should be consistent with a certain extent of system behavior and with a minimum of command and control. The window of allowed

activities should be defined and maintained with respect to the physical suitability of the area. Practical and managed river restoration measures can be applied to minimize impact.

3. Areas for ecosystem service generation. Natural ecosystem behavior and dynamics should be allowed in these areas. The evolution within these systems can serve as indicators for impact management of activities in type 1 areas. The challenge is to break with command and control and to restore a balance of ecosystem service generation and exploitation.

To enhance adaptive river basin management, there is a need for a conceptual framework that provides a thinking method that can be applied to assess various actions, activities and problems that relate to the basin. In this concept, human interactions with the water system should be placed centrally as we are part of the system. Metaphorically, we should act and react like ecosystems and strive for maximum efficiency and resource independence. To achieve efficient use of resources we need to improve internal recycling and re-use. Bundling and centralizing flows enhance management and control of interactions. By this means we have lower dependence of resources and lower impact on our environment. Likewise evolved ecosystems we need to attribute ourselves the attributes of a climax ecosystem. Economic and societal needs are driving forces for the use and exploitation of goods and services. The anthropogenic activities often unnecessarily interact and impact with the system. Activities can be conceptualized by interaction schemes, which are a schematics representation of the interactions between a stakeholder activity and the water system. These schemes are adapted from Tjallingii's guiding models (Tjallingii 1995; Tjallingii 2000) and represent the interactions, impacts and possible mitigation measures to minimize negative impact on the water system on different scale levels.

Hence we can attribute certain activities with optimal conditions in which interactions are minimized (e.g., agricultural production of crops that is conform the natural hydrological and topographical conditions). This is in contrast to the general style of management and control in which the characteristics of the waters system are adapted to the desired conditions by altering structures and processes of ecosystems. Maintaining the approach of command and control will eventually lead to degeneration of ecosystem service generation and consequently jeopardize generation of goods and services that are crucial to other economic and societal needs. Secondly this approach is unsustainable in terms of development, as growing pressures would lead to conflicts between societal functions and corresponding activities. As activities do not take place along gradients of suitability opposing objectives of land-use will ultimately result in conflicts with need excessive management efforts and costs to resolve. As natural gradients are not considered in land-use planning, there will be a larger friction line between the different functions. This friction line evenly increases if land-use is fragmented as the area/border ratio increases.

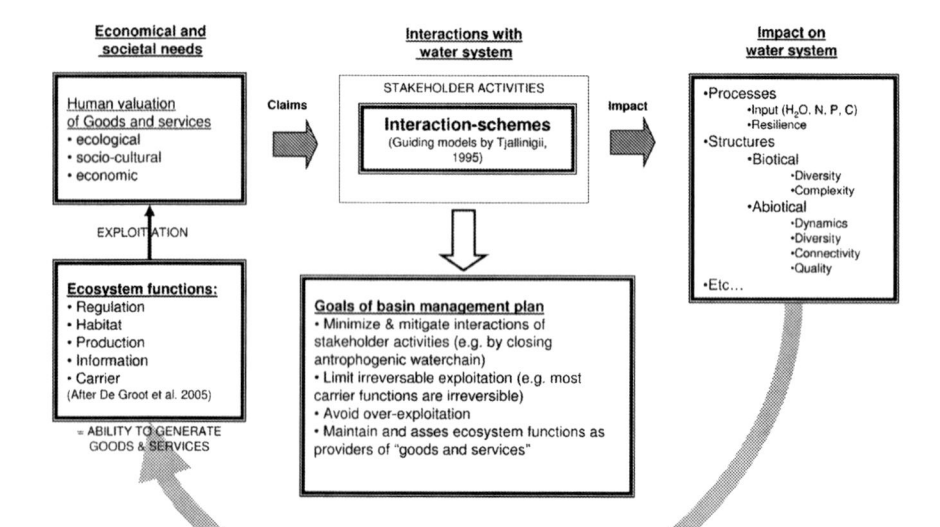

Figure 2. Conceptual scheme for basin management interactions

A river basin management plan should try to minimize interactions in general terms by minimizing:

- Dependency on the water system by efficient use of goods and services. Use fewer resources by efficient planning and technology. Closing of cycles by re-use and recycling.
- The extent of interactions between activities and the environment by applying large scale planning that complies with the natural characteristics and variability ranges. Minimize internal conflicts between adjacent functions and minimize fragmentation.

- Impact by reducing activities or minimizing impact per unit of activity: release impact on points where the system can absorb it with minimum impact, show regularity in impacts and allow system adaptation to improve resilience to these impacts.

Conditional to this approach, is the development of a general procedural framework (Fig. 3) that encapsulates aspects of information management, issue analysis, spatial analysis and public participation. The procedure must be adapted to enhance institutional and stakeholder cooperation and participation.

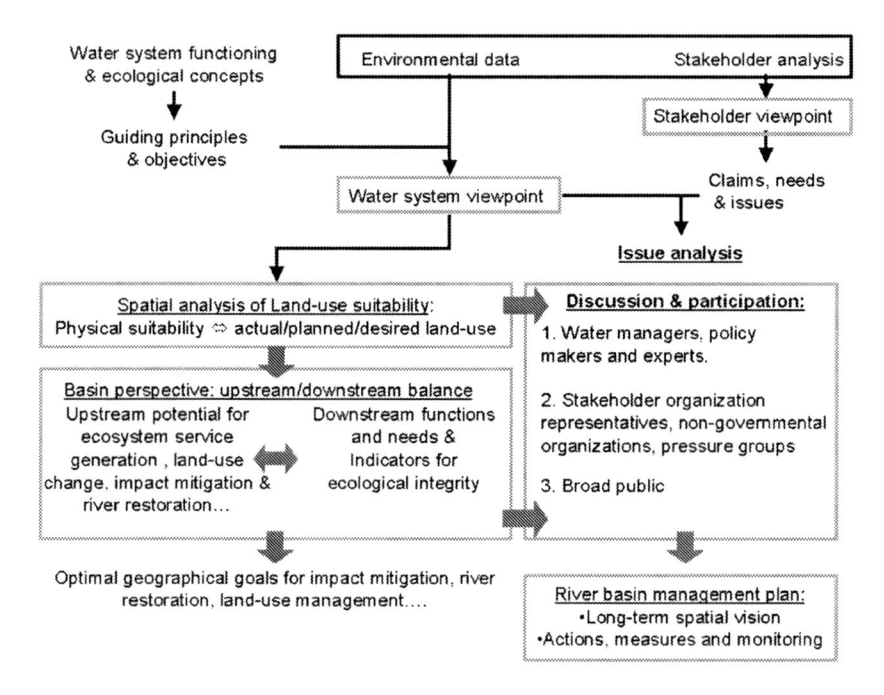

Figure 3. Methodological framework for elaboration of river basin management plans

A starting point would be an inventory of environmental data and a stakeholder analysis. Secondly, there should be agreement between and within institutional and stakeholder organizations on general principles regarding integrated water management. These guiding principles are used to translate and process the information into concrete measures and actions. Ideally, integrated water management should be linked to certain physical constraints that are clear and definable. The issue analysis addresses the problems and conflicts that are experienced by the stakeholder and water managers by analyzing the issues from a water system viewpoint. From this perspective the potential management and control measures first need to be assessed in terms of their impacts on the water system. Hence issues should always be traced towards their fundamental source of origin. The issue analysis also includes a spatial analysis tool that confronts actual, planned and desired land-use related activities with their respective physical suitability. The physical suitability should reflect an inverse gradation to the necessity of command and control to make the activity possible (or optimize its yields/efficiency). Hence, if there is no need for command and control for a specific activity at a specific location, consequently there is minimal impact and a high physical suitability. To maker this approach workable, we will have to make an abstraction of activities and the water system in terms of the range of variability. Possible criteria for the physical suitability are infiltration rate, infiltration capacity, erosion sensitivity, natural range of groundwater fluctuations and inundation probability and basin characterization (Strahler classification, River continuum concept, Stream zonation concept). On their turn these criteria are derived from combinations of available data. For example the inundation probability will be derived from the combination of multiple GIS-layers (landscape depressions, valley morphology, indicative historical deposition of sediments, reports of recent inundation and modeling results). The combination of and different information sources with different spatial scales, accuracy and credibility should allow an aggregated differentiation into indices of credibility and suitability. The approach can be applied to multiple types of land-use. From the perspective of the water system we need to analyze the physical and practical (actual situation) potency for active restoration of ecosystem service generation. We consider at least three major aspects of the hydrological cycle being infiltration, water conservation and flooding. They can serve as building units for the restoration of ecosystem generation within a basin perspective.

How do we define a sustainable balance of ecosystem service generation and ecosystem service exploitation/use? To provide a long-term sustainable use of the water system, there is a need to define limits to protect the carrying capacity to restore a balance at between the upstream supply of goods and services and the downstream use of goods and services (Agarwal, Braga et al., 1999). To answer these issues regarding the practical development of integrated decision support, there is a need for a simple and transparent model that can be easily applied and that can focus attention on the policy questions. The overall policy question is how to reach sustainability and acceptable ecological status with the least cost and maximal social acceptance. The model explores the need and the potential for measures and actions that can be taken. It is also a method to assess the importance of ecosystems in reducing the occurrence of environmental phenomena (flooding, anoxia, eutrophication, erosion etc...) in relation to its upstream input and downstream service area. These phenomena are to some extent mitigated in magnitude, frequency, recovery time and scale by ecosystem functioning. Crucial resilience factors on a basin scale are to be defined and depend on the characteristics of the upstream area and should therefore have sufficient structures and ecosystems in order to assimilate perturbations. The management challenge is to that extent to restore a natural and characteristic diversity of ecosystem and create semi-artificial opportunities for ecosystem development that can compensate for imbalances due to anthropogenic impact.

This method can be used to balance the need and benefits for additional measures in upstream areas with respect to their effect on mitigating the system vulnerability. It is by that way measures can be distributed spatially in a cost-effective manner. The potential for the implementation of instruments like land-management subsidies, land-use change/trading can be easily explored by the spatial analysis. The aim is to include four groups that are subject to geographical optimization and for which the actual situation and potential measures could be explored.

- Land-use related measures and actions.
    - Land-use change/trade.
    - Impact management/mitigation.
    - Ecosystem restoration/ecosystem service generation.
- River and riparian related measures and actions.
    - Agreements on functions and use of rivers and riparian zones.
    - Restoration of riparian zones for impact mitigation of adjacent land-use (buffer strips).
    - River restoration projects (meandering, gradient reconstruction).

- Sewage treatment plants investments.
  - Improve efficiency and nutrient removal.
  - Limit sewage system overflows.
  - Real time control of effluent release, sewage operation.
- Water demand management.
  - Reduce water needs by improved efficiency, re-cycling and re-use.
  - Reduce water allocation dependence by storage.

How do we define downstream needs? An initial condition would be ecological integrity as we are compelled to meet the good ecological status of our water bodies according to the EWFD. Also here we could make use of riverine ecology concepts and use them to define a referential status and a natural range of variability. A set of indicators should be developed to express the local (river segment) range of variability of the river characteristics (oxygenation, flow fluctuations, nutrient concentrations, water depth, turbidity, etc.). The indicators can be obtained from historical data, model predictions, monitoring data or can be estimated by upstream and local characteristics. A second set of conditions should be derived from the actual and desirable user-functions of watercourses and riparian zones. These functions can be derived from the actual user functions, adjacent land-use and by making inquiries. The functions are to be translated into certain ranges of river characteristics that are optimal-negative-unacceptable for that function. The indicated vulnerability might then conflict or not conflict with these user functions. It is possible that user functions are in conflict with each other and have non-overlapping requirements regarding the acceptable ranges of river characteristics (e.g., a sewage discharge point and water recreation are irreconcilable because of bacterial contamination hazard). It might therefore be appropriate to confront user functions within a certain spatial sphere and resolve eventual function allocation with the involved users. The described tools and concepts have to be embedded in formal planning procedures for river basin management. Participation and transparency is of vital importance as cooperation is requested of various policy domains and competence levels. Secondly the approach has to be supported by various organizations and interest groups. Finally there is a task to inform and convince the broad public using pilot projects, information sessions and educational mechanisms.

4. Conclusions

Although the term "integrated" has become increasingly popular, it is not always clear how this term is being applied. We still face a historical command and control approach to water management. Ecosystem functioning and ecology are often disregarded in hydro-technical engineering approaches. We should also be concerned about the use of numerical models and decision support tools as they still suffer many imperceptible drawbacks and misconceptions. They are typically focused on the known, predictable, measurable and controllable aspects of water management. The development of adequate monitoring schemes and basic research on water system functioning are primary needs to progress in the field of water system modeling. The role of ecosystems in providing goods and services is gaining both scientific and political attention. A precautionary course needs to be taken in managing our environment as seemingly increased benefits might only be temporal, benefit specific user groups or have dramatic effects on other aspects of the water system. An ecosystem approach for water management is a way to integrate different objectives and functions, including the 'hidden services' of the water system. No single approach or model should be used. A range of tools is needed, including expert judgement and conceptual models. The presented conceptual framework is a first step toward adaptive management as it provides a way of thinking in which the water system is of central importance. Translating this concept into practice will reduce the level of ambition. One reason is that there are aspects of water system functioning we will probably never understand. The presented approach allows reductionism in complexity but does not narrowly focus on what we do know, control and manage. It reckons with presumptions and hypotheses about ecosystem behavior and allows us to implement a cautious course in pursuing sustainability. The second reason is that a profound change in governmental institutions will probably not take place abruptly. There will remain a habitual tendency to control and manage aspects of our environment, as it is deep-rooted in society. We have spent hundreds of years trying to conquer the wilderness and the forces of nature, and abandoning this line of thought can be hard for those that are stricken by adaptive management. It is a reason to search and consolidate the chances and opportunities that exist today. This means not only restoration and impact mitigation but also putting a stop to further undesirable developments in land-use. Poor spatial planning and corresponding fragmented land-use requires an increasing level of command and control that will eventually lead to failure as control either fails in adequacy or efficiency. The value of spatial analysis and the balancing model is that it explores the needs and opportunities in a source-oriented manner. A balance between ecosystem service exploitation and generation on a basin scale means also that we have to consider the fluxes that occur over the system boundaries. The import of food, fodder, fertilizer, oil, water and other substances through controlled or uncontrolled transport can influence a watershed budget. If we speak about land-use changes since World War II, the location of agricultural production has shifted strongly toward generalization and scaling-up of cultivation (Bouma et al., 1998). Are we still able to manage on an eco-regional approach if the real pressures are macro-economic?

Acknowledgements

This work is inspired by the experiences, discussions and presentations throughout the six pilot study meetings on Integrated Water Management, sponsored by NATO – Committee on Challenges of the Modern Society. The geo-spatial, cultural and disciplinary diversity of the participants, has provided a unique opportunity to exchange expertise in water system research and moreover to learn from comparison by presenting examples to build upon.

References

Agarwal, A., B. Braga, et al., (1999). Proceedings: Towards Upstream/Downstream Hydrosolidarity A SIWI/IWRA Seminar, Stockholm, 14 August 1999. SIWI/IWRA Seminar, Stockholm, Stockholm International Water Institute, Siwi.
Agee, J. K. and D. I. L. Johnson (1988). "Ecosystem management for parks and wilderness." University of Washington Press, Seattle, Washington.

Allan, J. D., D. L. Erickson, et al., (1997). "The influence of catchment land use on stream integrity across multiple spatial scales." Freshwater Biology 37(1)): 149–161.

Allen, T. F. H. and T. B. Starr (1982). "Hierarchy." University of Chicago Press, Chicago.

Amoros, C. A., L. Roux, et al. (1987). "A method for applied ecological studies of fluvial hydrosystems." Regulated Rivers Research & Management 1: 17–38.

Antrop, M. and V. Van Eetvelde (2000). "Holistic aspects of suburban landscapes: visual image interpretation and landscape metrics." Landscape and Urban Planning 50(1–3): 43–58.

Baker, W. L. (1995). "Long term response of disturbance landscapes to human intervention and global change." Landscape Ecology, 10, 143–159.

Baskin, Y. (1997). "The Work of Nature: How the Diversity of Life Sustains Us." Washington D.C., Covelo, CA: Island Press.

Beck, M. B. (2005). "Vulnerability of water quality in intensively developing urban watersheds." Environmental Modelling & Software 20(4): 381–400.

Boon, P. J. (1992). Essential elements in the case for river restoration. River Conservation and Management. P. J. Boon, P. Calow and G. E. Petts. Chichester, UK, John Wiley & Sons Ltd: pp. 10–33.

Born, S. M. and W. C. Sonzogni (1995). "Integrated Environmental-Management – Strengthening the Conceptualization." Environmental Management 19(2): 167–181.

Bouma, J. (2002). "Land quality indicators of sustainable land management across scales." Agriculture Ecosystems & Environment 88(2): 129–136.

Bouma, J. (2006). "Hydropedology as a powerful tool for environmental policy research." Geoderma 131(3–4): 275–286.

Briggs, S. (2003). "Command and control in natural resource management: Revisiting Holling and Meffe." Ecological Management and Restoration 4(3): 161–162.

Callicot, J. B. and K. Mumford (1997). "Ecological sustainability as a conservation concept." Conservation Biology 11(1): 32–40.

Carpenter, S. R., W. A. Brock, et al. (1999). "Ecological and social dynamics in simple models of ecosystem management." Conservation Ecology 3(1).

Clark W. and Munn R, e. (1986). "Sustainable Development of the Biosphere, W." Clark and R. Munn, eds. Cambridge: Cambridge University Press, 1986.

Cockx, J. (1996). First National Report of Belgium to the Convention on Biological Diversity. Brussels, Environment, Nature, Land and Water Management Administration (AMINAL) for Belgian National Focal Point to the Convention on Biological Diversity – Royal Belgian Institute of Natural Sciences.

Costanza, R., R. d'Arge, et al. (1997). "The value of the world's ecosystem services and natural capital. Nature" 387(6630): 253–260.

Costanza, R. and M. Mageau (1999). "What is a healthy ecosystem?" Aquatic Ecology(33): 105–115.

Daily, G. (1997). "Nature's Services: Societal Dependence on Natural Ecosystems." Island Press, Washington, DC.

De Groot, R. (2005). "Function-analysis and valuation as a tool to assess land use conflicts in planning for sustainable, multi-functional landscapes." Landscape and Urban Planning In Press, Corrected Proof.

De Groot, R. S., M. A. Wilson, et al. (2002). "A typology for the classification, description and valuation of ecosystem functions, goods and services." Ecological Economics 41(3): 393–408.

Doran, J. W. and M. R. Zeiss (2000). "Soil health and sustainability: managing the biotic component of soil quality." Applied Soil Ecology 15(1): 3–11.

Downs, P. W., K. J. Gregory, et al. (1991). "How Integrated Is River Basin Management." Environmental Management 15(3): 299–309.

Dumanski, J. and C. Pieri (2000). "Land quality indicators: research plan." Agriculture Ecosystems & Environment 81(2): 93–102.

Elwood, J. W., J. D. Newbold, et al. (1983). "Resource spiralling: An opearational paradigm for analyzing lotic systems." In T.D. Fontaine III & S.M. Bartell eds. Dynamics of lotic ecosystems. Ann Arbor, MI, USA, Ann Arbor Science Publishers. 3–27.

Enderlein, R. and F. Bernardini (2005). "Nature for water: Ecosystem services and water management." Natural Resources Forum 29(3): 253–255.

Everard, M. (2004). "Investing in sustainable catchments." Science of The Total Environment 324(1–3): 1–24.

Fiedler, P. I., P. S. White, et al. (1997). The paradigm shift in ecology and its implications for conservation. The Ecological Basis of Conservation: Heterogeneity, Ecosystems, and Biodiversity. S. T. A. Pickett, R. S. Ostfeld, M. Shachak and G. E. Likens. New York, Chapman and Hall: 83–92.

Fisher, S. G., N. B. Grimm, et al. (1998). "Material Spiraling in Stream Corridors: A Telescoping Ecosystem Model." Ecosystems 1(1): 1934.

Frissell, C. A., W. J. Liss, et al. (1986). "A hierarchical framework for stream habitat classification: Viewing streams in a watershed context." Environmental Management, 10: 199–214.

Gardiner, J. L. (1984). "Sustainable development for river catchment." Journal of the Chartered Institution of Water and Environmental Management (8 june): 308–319.

Gibbons, M., C. Limoges, et al. (1994). "The new production of knowledge: The dynamics of science and research in contemporary societies." London: Sage.

Grigg, N. S. (1996). "Water resources management: principles, regulations and cases." ISBN 0-07-024782-X.

Grumbine, R. E. (1994). "What Is Ecosystem Management?" Conservation Biology 8(1): 27–38.

Grumbine, R. E. (1997). "Reflections on "What is Ecosystem Management?"" Conservation Biology 11(1): 41–47.

Gulinck, H. and T. Wagendorp (2002). "References for fragmentation analysis of the rural matrix in cultural landscapes." Landscape and Urban Planning 58(2–4): 137–146.

Gustavson, K. R., S. C. Lonergan, et al. (1999). "Selection and modeling of sustainable development indicators: a case study of the Fraser River Basin, British Columbia." Ecological Economics 28(1): 117–132.

Hansson, L.-A., C. Bronmark, et al. (2005). "Conflicting demands on wetland ecosystem services: nutrient retention, biodiversity or both?" Freshwater Biology 50(4): 705–714.

Harremoës, P. (2000). "Scientific incertitude in environmental analysis and decision making." from http://www.knaw.nl/heinekenprizes/pdf/11.pdf.

Harremoës, P. (2003). "The Need to Account for Uncertainty in Public Decision Making Related to Technological Change." Integrated Assessment 4(No 1).

Harremoes, P. and H. Madsen (1999). "Fiction and reality in the modeling world – Balance between simplicity and complexity, calibration and identifiability, verification and falsification." Water Science and Technology 39(9): 1–8.

Herath, H. S. B., Park, Chan S. (2001). "Real Options Valuation And Its Relationship To Bayesian Decision-Making Methods." Engineering Economist. 46(Issue 1).

Hobbs, R. J. and J. A. Harris (2001). "Restoration Ecology: Repairing the Earth's Ecosystems in the New Millennium." Restoration Ecology 9(2): 239–246.

Hobbs, R. J., D. A. Saunders, et al. (1993). Reintegrating fragmented landscapes: towards sustainable production and nature conservation. Changes in biota. R. J. Hobbs and D. A. Saunders. New York, Springer: 65–106.

Holling, C. S. (1992). "Cross-scale morphology, geometry and dynamics of ecosystems." Ecological Monographs 62(4): 447–502.

Holling, C. S. and G. K. Meffe (1996). "Command and control and the pathology of natural resource management." Conservation Biology 10(2): 328–337.

Hosseini, H. U. (2001). "Uncertainty and perceptual problems causing government failures in less advanced nations." Journal of SocioEconomics 30(3): 263–271.

Hutchings, C., J. Struve, et al. (2002). State of the Art Review (state of the art in linking models). IT Frameworks (HarmonIT) Contract EVK1-CT-2001-00090 C. Hutchings, HR Wallingford Report SR 598.

Illies, J. and L. Botosaneanu (1963). "Problemes et methodes de la classification et de la zonation ecologiques des eaux courantes, consideres surtout du point de vue faunistique." Mitt. Internat. Verein. Limnology 12: 1–57.

Insley, M. (2002). "A Real Options Approach to the Valuation of a Forestry Investment." Journal of Environmental Economics and Management 44(3): 471–492.

IWRA, A. Agarwal, et al. (1999). PROCEEDINGS: Towards upstream/downstream hydrosolidarity. SIWI/IWRA SEMINAR, STOCKHOLM, Stockholm International Water Institute, SIWI

Jakeman, A. J. and R. A. Letcher (2003). "Integrated assessment and modeling: features, principles and examples for catchment management." Environmental Modelling & Software 18(6): 491–501.

Jessel, B. and J. Jacobs (2005). "Land use scenario development and stakeholder involvement as tools for watershed management within the Havel River Basin." Limnologica – Ecology and Management of Inland Waters 35(3): 220–233.

Jewitt, G. (2002). "Can Integrated Water Resources Management sustain the provision of ecosystem goods and services?" Physics and Chemistry of the Earth, Parts A/B/C 27(11–22): 887–895.

Junk, W. J. and K. M. Wantzen (2003). The flood pulse concept: new aspects, approaches and applications – an update. Second International Symposium on the Management of Large Rivers for Fisheries, Phnom Penh, Cambodia, Food and Agriculture Organization and Mekong River Commission, FAO Regional Office for Asia and the Pacific.

Karlen, D. L., M. J. Mausbach, et al. (1997). "Soil quality: A concept, definition, and framework for evaluation." Soil Science Society of America Journal 61(1): 4–10.

Kassar, I. and P. Lasserre (2004). "Species preservation and biodiversity value: a real options approach." Journal of Environmental Economics and Management 48(2): 857–879.

Klotzli, F. and A. P. Grootjans (2001). "Restoration of Natural and Semi-Natural Wetland Systems in Central Europe: Progress and Predictability of Developments." Restoration Ecology 9(2): 209–219.

Levin, S. A. (1992). "The problem of pattern and scale in ecology." Ecology 73:1943–1967.

Levin, S. A. (1998). "Ecosystems and the biosphere as complex adaptive systems." Ecosystems 1: 431–436.

Levin, S. A. (1999). "Fragile dominion: complexity and the commons." Perseus Books, Reading, Massachusetts, USA.

Levin, S. A. (1999). "Towards a science of ecological management." Conservation Ecology 3 2: 6. [online] URL: http://www.consecol.org/3/iss2/art6/ .

Light, S., L. Gunderson, et al. (1995). "The Everglades: Evolution of Management in a Turbulent Ecosystem." In Barriers and Bridges to the Renewal of Ecosystems and Institutions, eds. L. Gunderson, C. Holling, and S. Light. New York: Columbia University Press.

Ludwig, D., R. Hilborn, et al. (1993). "Uncertainty, Resource Exploitation, and Conservation: Lessons from History." Science 260(17): 36.

Lundqvist, J. and M. Falkenmark (2000). "Drainage basin morphology: a starting point for balancing water needs, land use and fishery protection." Fisheries Management and Ecology 7(1–2): 1–14.

McClain, M. E., E. W. Boyer, et al. (2003). "Biogeochemical Hot Spots and Hot Moments at the Interface of Terrestrial and Aquatic Ecosystems." Ecosystems 6(4): 301–312.

Meffe, G. K. and C. R. Carroll (1997). "Conservation Biology, 2nd edition." Sinauer Associates, Inc. Sunderland, Massachusetts.

Merriam, G. (1984). Connectivity: a fundamental ecological characteristic of landscape pattern. Proceedings of the First International Seminar on Methodology in Landscape Ecological Research and Planning: Theme I: Landscape & Ecological Concepts, Roskilde, Denmark., Roskilde University Centre Book Company, .

Naiman, R. J. and H. Decamps (1990). "The ecology and management of aquatic-terrestrial ecotones." UNESCO MAB series. The Parthenon Publishing Group. UNESCO Paris. Paris, France, 316 pp.

Newbold, J. D., J. W. Elwood, et al. (1981). "Measuring nutrient spiralling in streams." Canadian Journal of Fisheries and Aquatic Science(38): 860–863.

O'Neil, R. V., A. R. Johnson, et al. (1989). "A hierarchical framework for the analysis of scale." Landscape Ecology, 3: 193–205.

Odum, E. P. (1971). Fundamentals of ecology. Philadelphia., W.B. Saunders Co., .

Pallottino, S., G. M. Sechi, et al. (2005). "A DSS for water resources management under uncertainty by scenario analysis." Environmental Modelling & Software 20(8): 1031–1042.

Parker, P., R. Letcher, et al. (2002). "Progress in integrated assessment and modeling." Environmental Modelling & Software 17(3): 209–217.

Pavlikakis, G. E. and V. A. Tsihrintzis (2000). "Ecosystem Management: A review of a new concept and methodology." Water Resources Management 14(4): 257–283.

Petersen, R., L. M. Petersen, et al. (1990). A Building-block model for stream restoration. River conservation and management P. J. Boon, P. Calow and G. Petts. Chichester, United Kingdom, John Wiley and Sons Ltd: 293–310.

Peterson, G. (2000). "Scaling ecoligical dynamics: self-organization, hierarchical structure, and ecological resilience." Climatic Change(44): 291–309.

Peterson, G., C. R. Allen, et al. (1998). "Ecological Resilience, Biodiversity, and Scale." Ecosystems(1): 6–18.

Petts, G. E. (1989). Historical Changes of Large Alluvial Rivers in Western Europe. Chichester., Wiley,

Petts, G. E. and C. E. Amoros (1996). "Fluvial Hydrosystems." Chapman & Hall, London.

Pickett, S. (1997). The Ecological Basis of Conservation: Heterogeneity, Ecosystems, and Biodiversity. New York, Chapman and Hall.

Pickett, S., V. T. Parker, et al. (1992). The new paradigm in ecology: Implications for conservation biology above the species level. Conservation Biology: The Theory and Practice of Nature Conservation, preservation, and Management. P. L. Fiedler and S. K. Jain. New York, Chapman and Hall: 65–88.

Pinay, G., H. Decamps, et al. (1999). "The spiraling concept and nitrogen cycling in large river floodplain." Archiv. Für Hydrobiologie, 143: 281–291.

Pindyck, R. S. (2002). "Optimal timing problems in environmental economics." Journal of Economic Dynamics and Control 26(9–10): 1677–1697.

Quevauviller, P., P. Balabanis, et al. (2005). "Science-policy integration needs in support of the implementation of the EU Water Framework Directive." Environmental Science & Policy 8(3): 203–211.

Radwan, M., P. Willems, et al. (2003). "Modelling of dissolved oxygen and biochemical oxygen demand in river water using a detailed and a simplified model." Journal of River Basin Management. 1(No. 2):. 97–103.

Rapport, D. J. (1992). "Evaluating ecosystem health." Journal of Aquatic Ecosystem Health, 1, 15–24.

Reitsma, R. F. and J. C. Carron (1997). " Object-oriented Simulation and Evaluation of River Basin Operations," Journal of Geographic Information and Decision Analysis 1(1): 9–24.

Ripl, W. (2003). "Water: the bloodstream of the biosphere." Philosophical Transactions of the Royal Society of London Series B-Biological Sciences 358(1440): 1921–1934.

Ripl, W., Hildmann, Ch., Janssen, T., Gerlach, I., Heller, S. & Ridgill, S. (1995). Sustainable redevelopment of a river and its catchment – the Stör River Project. Restoration of Stream Ecosystems – an integrated catchment approach. M. B. Eiseltová, J., IWRB Publication. 37: 76–112

Schatzki, T. U. (2003). "Options, uncertainty and sunk costs:: an empirical analysis of land use change." Journal of Environmental Economics and Management 46(1): 86–105.

Scheffer, M., W. Brock, et al. (2000). "Socioeconomic Mechanisms Preventing Optimum Use of Ecosystem Services: An Interdisciplinary theoretical Analysis." Ecosystems 3: 451–471.

Scheffer, M. and S. R. Carpenter (2003). "Catastrophic regime shifts in ecosystems: linking theory to observation." Trends in Ecology & Evolution 18(12): 648–656.

Scheffer, M. C., S Foley, JA Folke, C Walker, B (2001). "Catastrophic shifts in ecosystems." NATURE 413(11 oct 2001): 591–596.

Schroder, W., G. Schmidt, et al. (2003). "Spatial representativity and methodical comparability of data and sites of soil monitoring." Journal of Plant Nutrition and Soil Science-Zeitschrift Fur Pflanzenernahrung Und Bodenkunde 166(5): 649–659.

Simonovic, S. P. (2000). "Tools for Water Management: One View of the Future." Water International, (IWRA) International Water Resources Association 25(1): 76–88.

Stanford Ja, W. J. V. (1993). "An ecosystem perpective of alluvial rivers: connectivity and the hyporheic corridor." Journal of the North American Benthological Society 12, 48–60.

Strahler, A. N. (1957). "Quantitative analysis of watershed geomorphology." Trans. Am. Geophys. Union 38: 913–920.

Talbot, L. (1996). "Living resource conservation: An international overview." Marine Mammal Commission, Washington, D.C. 56pp.

Tjallingii, S. P. (1995). Ecopolis: Strategies for ecologically sound urban development. Leiden, Backhuys Publishers.

Tjallingii, S. P. (2000). "Ecology on the edge: Landscape and ecology between town and country." Landscape and Urban Planning 48(3–4): 103–119.

Tockner K, S. F. and J. V. Ward (1998). "Conservation by restoration: the management concept for a river-floodplain system on the Danube River in Austria." Aquatic Conservation, 47, 7186.

Tockner, K., J. V. Ward, et al. (2002). "Riverine landscapes: an introduction." Freshwater Biology 47(4): 497–500.

Ulanowicz, R. E. (1986). "Growth and development: ecosystem phenomenology." Springer, New York.

Upadhyay, R. K. and V. K. Rai (1997). "Why chaos is rarely observed in natural populations." Chaos, Soliton and Fractals, 8 12:1933–1939.

Van Eetvelde, V. and M. Antrop (2005). "The significance of landscape relic zones in relation to soil conditions, settlement pattern and territories in Flanders." Landscape and Urban Planning 70(1–2): 127–141.

Vannote, R. L., G. W. Minshall, et al. (1980). "The River Continuum Concept." Can. J. Fish. Aquat. Sci. 37: 130–137.

Vanrolleghem, P. A., L. Benedetti, et al. (2005). "Modelling and real-time control of the integrated urban wastewater system." Environmental Modelling & Software 20(4): 427–442.

VIWC (2000). Ontwerp Waterbeleidsplan Vlaanderen 2002 – 2006. Brussel, Departement Leefmilieu en Infrastructuur: 242 pp.

Walker, A. W. E., P. Harremoes, et al. (2003). "Defining Uncertainty: A Conceptual Basis for Uncertainty Management in Model-Based Decision Support." Integrated Assessment 4(No 1).

Walker, W. E. and V. A. W. J. Marchau (2003). "Dealing With Uncertainty in Policy Analysis and Policymaking-Introduction to Papers." Integrated Assessment 4 (No 1).

Ward, J. V. (1998). "Riverine landscapes: biodiveristy patterns, disturbance regimes, and aquatic conservation." Biological Conservation 83, 3: 269–278.
Ward, J. V. and J. A. Stanford (1983). "The serial discontinuity concept of lotic ecosystems." Dynamics of lotic ecosystems. T. D. Fontaine and S. M. Bartell. Ann Arbor, Ann Arbor Science: 29–42.
Ward, J. V. and J. A. Stanford (1995). "The serial discontinuity concept: extending the model to floodplain rivers." Regulated Rivers: Research and Management 10: 159–168.
Ward, J. V., K. Tockner, et al. (2002). "Riverine landscape diversity." Freshwater Biology 47(4): 517–539.
Ward, J. V., K. Tockner, et al. (1999). "Biodiversity of floodplain ecosystems:ecotones and connectivity." Regulated Rivers: Research and Management 15:125–139.
Ward, J. V. and J. A. Wiens (2001). "Ecotones of riverine ecosystems: Role and typology, spatio-temporal dynamics, and river regulation." Ecohydrology and Hydrobiology(1): 25–36.
Wasson, J.-G., M.-H. Tusseau-Vuillemin, et al. (2003). "What kind of water models are needed for the implementation of the European Water Framework Directive: Examples from France." Intl. J. River Basin Management 1(No. 2): pp. 125–135.
Welp, M. (2001). "The use of decision support tools in participatory river basin management." Physics and Chemistry of the Earth, Part B: Hydrology, Oceans and Atmosphere 26(7–8): 535–539.
Wester, P., D. J. Merrey, et al. (2003). "Boundaries of Consent: Stakeholder Representation in River Basin Management in Mexico and South Africa." World Development 31(5): 797–812.
Wurbs, R. A. (2005). "Modeling river/reservoir system management, water allocation, and supply reliability." Journal of Hydrology 300(1–4): 100–113.
Wynne, B. (1992). "Uncertainty and Environmental Learning." Global Environmental Change, Butterworth-Heinemann Ltd. pp. 111–127.
Yaffee, S. L. (1999). "Three Faces of Ecosystem Management." Conservation Biology 13(4): 713–725.
Zalewski, M., B. Bis, et al. (2001). "Riparian ecotone as a key factor for stream restoration." Ecohydrology & Hydrobiology 1 1–2: 245–253.
Zalewski, M., W. Puchalski, et al. (1994). "Riparian ecotones and fish communities in rivers – intermediate complexity hypothesis." 152–160 In I.G. Cowx Ed., Rehabilitation of Freshwater Fisheries. Fishing News Book.

PART IV: WORKSHOP CONCLUSIONS AND RECOMMENDATIONS

ARW CONCLUSIONS AND RECOMMENDATIONS

The materials presented at this Advanced Research Workshop provided an overview of transboundary water management problems in Central Asia, North America, Europe, and the Caucuses. The discussions at this workshop were focused on the legal, technical, and institutional aspects of transboundary water management problems as well as on potential tools and approaches to deal with transboundary water management issues. During the last day of the Workshop, there was a special session on the development of primary conclusions, recommendations, and the identification of additional areas for research, training, and education in the area of transboundary water management. The discussions highlighted the following important issues that need to be addressed:

1. Legal Tools. Water resource management in Central Asia is complex, involving many competing jurisdictions and water users (e.g., industrial users, agricultural users, power-generating users, domestic users), and will therefore require broad-based cooperation by all affected countries. Effective long-term region-wide management will require sacrifice and will also require a long-term view for complete implementation of effective strategies to manage this regional resource. More specific recommendations are the following:

- There is a need for the development of a legal/regulatory framework, rules, and mechanisms of Integrated Water Resource Management (IWRM) in transboundary water basins in Central Asia.

- There is a need for a regionally accepted approach for addressing water-related conflicts/disputes/coordination problems.

- Examples of 'successful' institutional structures in the world for addressing transboundary water issues need to be identified and described.

- Examples of transboundary water conflicts worldwide, how they were addressed and what factors lead to both success and failure should also be disseminated.

- A regional commission that meets regularly and has authority to make decisions, impose penalties and have open access to data and countries be established.

2. Economic Analysis/Mechanisms. Economic mechanisms (e.g., pricing, taxation, compensation approaches) for improved water resource management ought to be considered and incorporated by policy makers in Central Asia. These mechanisms should be clearly configured as both incentives and disincentives, as needed, to attain the sustainable environmental goals and objectives of the region. More specific recommendations are the following:

- Uniform concepts of the economic, environmental, and other social aspects of shared water use in the CA transboundary rivers be established.
- Develop a methodology for integrated technical and economic assessments, including socioeconomic evaluations of damages and benefits of shared use of transboundary water resources.
- Water and hydro-power energy issues be integrated when evaluating the environment in the Central Asia Region.
- An inventory of vulnerable hydro-technical infrastructures (e.g., dams, reservoirs, hydro-electric plants) be completed in order to establish risk-based counter measures and mitigation schemes from man-made and natural threats.

3. Ecologically Sustainable Flows Approach. In water resources planning in Central Asia, insufficient attention has been given to the ecological and social consequences (e.g., ecological sustainability, impacts on fish and wildlife, human health impacts) of inefficient water use decisions. Emphasis placed primarily on the immediate engineering and production benefits of water resource use neglects important attributes of water (e.g., ecological, human health) as a fundamental resource throughout the region. It is recommended that:

- A uniformly accepted concept of IWRM for Central Asia be developed based on an ecosystem wide approach.
- A pilot project be implemented to develop a methodology and practical application to determine ecologically sustainable

environmental flows of the transboundary rivers/aquifers in Central Asia.

4. Integrated Water Resources Management (IWRM). There is a lack of a consistent understanding of the concept of IWRM by decision-makers and the general public in Central Asia. This lack of knowledge will greatly hamper long-term, efficient and sustainable water resource management in Central Asia. It is recommended that:

- Integrated models be further developed and used in support of regional discussion, environmental, decision-making and vision development for IWRM.

- A NATO Science for Peace project (or similar project) on the use of 'return' waters in Central Asia be conducted. There is a need to incorporate the management of 'return waters'/ drainage waters in Integrated Water Resources Management (IWRM) approaches in Central Asia.

- Groundwater issues need to be fully integrated into 'integrated water management' approaches in Central Asia. Groundwater resources should be incorporated with surface water in regional river basin management strategies. (There are important shared aquifer systems in Central Asia.)

- Sediment transport to receiving water bodies in Central Asia be incorporated into IWRM modeling.

- The application of inefficient agricultural practices be evaluated in terms of their impacts on IWRM.

5. Informational Support. Efforts should be made to implement improved monitoring methodologies as part of a standardized approach that would involve the regional sharing of data from monitoring activities. These improved monitoring methodologies and data sharing activities would require the implementation of uniform standards throughout the Central Asian region for sample collection, data analyses, etc. It is recommended that:

- Improved informational support of transboundary water resource management in Central Asia be implemented, thereby providing data of higher quality.

- A regional web site (or clearinghouse) for storing and sharing regional water data collected by all entities throughout Aral Sea Basin be implemented.

- New environmental standards for water quality of the transboundary rivers in the Central Asian region be drafted for review and implementation.

- The efforts of scientists from the South Caucasus region and the Central Asian region be coordinated on transboundary water issues. Such an effort would facilitate exchanges of technical data on shared inter-regional water issues (via the Caspian Sea). This combined effort could be used to conduct comprehensive research on the pollution effects of the Caspian Sea by South Caucasus and Central Asian rivers.

6. Human Health Aspects of Transboundary Water Issues. It is recommended that:

- A comprehensive research program for determining the adverse public health effects of poor water quality be implemented in the Central Asian Region.

- Greater dissemination of information on new technologies for wastewater disposal be implemented.

7. Education and Training. Increased education and technical training for future water professionals, elected officials and other interested individuals are needed for improved transboundary water management throughout the region. It is recommended that:

- Training on Integrated Water Resource Management should be provided to young environmental professionals in the region.

- Educational training curricula and associated materials for mid-level ministry employees be developed.

- Dissemination of the information on completed TBW management projects by the United States Agency for International Development (USAID) and the European Union's Technical Aid to the Commonwealth of Independent States (TACIS) Program be expanded in order to further progress toward the efficient and sustainable utilization of water resources throughout the region.

Also, it was concluded that:

- In water resources planning in Central Asia, insufficient attention has been given to the social consequences of inefficient water use decisions. Emphasis primarily on the immediate engineering and production benefits of water resource use neglects important attributes of water as a fundamental resource throughout the region.

- There is a need to de-emphasize *problem identification* and move toward *cooperation and solutions* to problems. Efforts to encourage or require regional or multi-state cooperation in the development of water resource management strategies will be necessary for the efficient utilization of this shared common property resource.

SUBJECT INDEX

A
Amu Darya River, 16, 18, 49, 93, 106-107, 111-114, 250
Apalachicola-Chattahoochee-Flint basin, 195
aquatic ecosystems, 44
aquifers, 243, 244
Aral Sea, 33
Aral Sea Basin, 29, 79, 105
Armenia, 162
ARW, 1
Azerbaijan, 162

B
Basin Management Organizations (BVO), 31, 32
Basin Water Authority, 84
Basin Water Organizations, 85
basin-level management, 29
basin-wide vision process, 36
biodiversity, 283

C
carrying capacity (CC), 275
Caspian Sea, 6, 45, 48, 155, 171, 308
Caucasus, 43, 153, 164, 165, 171, 308
CCMS, 66
Central Asia, 225
Chu River, 133
command and control management, 267
compensation schemes for sharing water, 5
complexity paradigm, 269
conflict avoidance, 1
conflict resolution, 7, 65, 68, 73, 259

D
decision making process, 35, 65, 108, 135
decision support system (DSS), 273, 276

E
economic mechanisms, 24, 57, 58, 60, 143
ecosystem approach, 52, 73, 230, 293
ecosystem/watershed management, 266
environmental decision-making, 2
environmental flows, 224, 234, 239, 240
environmentally sustainable water management, 223
epidemiological analysis, 191
equitable and sustainable management practices, 244
equitable apportionment, 210, 214, 219, 220
equitable utilization principle, 195

F
Fergana Valley, 89, 91
Florida State University (FSU), 2
freshwater ecosystems, 223, 224

G
Georgia, 163
GIS-based assessment tool, 183
groundwater, 113

H
hazardous pollutants, 172, 173
heavy metals, 21, 155, 166, 174, 180
hydroelectric potential, 16
hydrologic cycle, 223, 224

I
ICWC, 32, 57, 102, 144, 255
IFAS, 86
IICER, 1
instream flow, 234
integrated assessment and modeling (IAM), 268, 272
integrated catchment management (ICM), 278
Integrated Water Management (IWM), 66, 264

311

SUBJECT INDEX

Integrated Water Resources Management (IWRM), 4, 40, 87, 108, 123, 224, 239, 243
integrated/holistic water management, 229
International basin, 155
International Association of Hydrologists (IAH), 247
international scale of water management, 265
Internationally Shared Aquifer Resources Management (ISARM), 247
Interstate Coordination Water Commission (ICWC), 32, 57, 102, 144, 255
irrigated agriculture, 11, 16, 30, 71, 172, 216
irrigation, 46
IWP, 2
IWRM, 87

K
Kashkadarya River, 14, 18
Kura-Araks, 171
Kura-Araks basin, 156, 164, 171, 181
Kyrgyzstan, 141

L
Lebap Province, 105
legal issues, 195
legal/technical/institutional determinants of water issues, 1

N
NATO, 65
NATO Advanced Research Workshop, 1, 6, 67, 74
NATO projects, 164
natural streamflow, 231
natural systems, 228, 266, 275
negotiations, 195, 200-202, 204, 206, 230

O
OSCE projects, 134

P
Pesticides, 21, 156, 180
potable water sources, 62, 183

potable water supply, 22
public health, 167
public participation, 58

R
radionuclides, 177
regional security, 6
return water, 22, 48, 52, 90, 95, 98, 103
risk assessment, 4
river basin commissions, 200
river basin management, 39, 288
river basin management plan, 74, 288
River Basin Water-Management Organization (BWO), 102

S
Science for Peace projects, 153
SIC, 32
social and technical infrastructure, 154, 165
soil erosion, 46
soil salinization, 63, 96
South Caucasus region, 155, 171, 308
spatial analysis, 185, 289, 294
STCU, 191
Sustainability, 223
sustainable development, 87
Syr Darya, 96
system wide management, 198

T
Tajikistan, 87
The Worldwide Hydrogeological Mapping and Assessment Programme, 250
Talas River, 128, 129
total water management (TWM), 258
transboundary aquifers, 243
transboundary rivers, 183
transboundary water management, 162
transboundary water resources, 67
transnational watercourses, 195
Turkmenistan, 105

U
UNESCO-IHE, 40, 41
uncertainty paradigm, 269
United Nations, 44, 161

SUBJECT INDEX

United States, 195
United States Supreme Court, 6, 213
Uzbekistan, 24

W
water allocation, 223
water flow formation, 11
water management capacity development, 29
water markets, 5
water modeling, 173
water monitoring, 155, 167
water pricing, 5
water quality, 1
water quality data, 174
water quality management, 26, 180, 184
water resources management, 268
water salinization, 18
water system theory, 278
water users associations, 88
waterbourne diseases, 183
Watercourse Convention, 221

Z
Zeravshan River, 14, 183

Printed in the United Kingdom
by Lightning Source UK Ltd.
124875UK00002B/103-120/A